Save Today's Environment *for* Many Morrows' Development

Muktipada Das

Originals

Delhi

First Published 2013

ISBN 13: 978-81-8454-136-6

ISBN 10: 81-8454-136-8

Published By

Originals

A-6, Second Floor, Nimri Commercial Centre,

Ashok Vihar Phase-4, Delhi - 110 052 (India)

Tel.: +91 11 2730 2453 Fax: 011 4706 1936

website: www.Lppindia.com

e-mail: info@Lppindia.com

Printed At

D K Fine Art Press P Ltd., Delhi-110052

Ph.: 2730 2451

PRINTED IN INDIA

Preface

I went in for research in Geology after Master's in Geology. But, unfortunately, I couldn't complete the research for Ph.D. degree and joined Coal India as a geologist. In the backdrop of frustrated venture of science research I desired to cultivate science in some other way and tried my hand in writing on popular science. Initially I could write a few articles in Bengali popular science magazines. But afterwards, owing to much pre-occupation, pursuing it further was not possible. Now, at this juncture of retirement, I have once again invoked that attempt.

The love for nature grew in my mind in course of my profession in a natural science like geology. Later in Coal India career, by virtue of being engaged in environmental management of coal mines and coalfields I could learn a lot about environment management. I then acquired a little bit of formal education on Industrial Pollution & Control through distant education channel. At this phase I developed desire to write a book on everyman's environmental science. The urge was enhanced on seeing conspicuous lack of concern and discipline among people about environment protection with rush for development and wealth.

My present effort is, therefore, to contribute something in dissemination of environmental awareness, concern and discipline among people towards the goal of sustainable development. While doing so I have tried my best to accommodate in this book the two-in-one objectives of campaign for environmental awareness, concern and discipline among the public at large as well as making it useful as a reference book for students in general. The presentation and language of the book have been maintained as clear and pleasant as possible. In doing so, barring the chapter titles, no subtitle/ heading has been used within a chapter for the sake of smooth reading without break. But I do not know how far I have been successful in this direction. That is left to be assessed by the readers. I have proposed a slogan to be raised for public awareness and have chosen the same for the title of the book as well.

Muktipada Das

e-mail: mukti.das256@gmail.com

Contents

List of Plates:

List of Figures:

List of Charts:

1

The Biosphere, Ecosystems & Ecology

Throughout the universe, life is known to exist only on one planet, the Earth, at least as of now. For, as we know, the essential requirements of life process are available in an orderly manner on earth alone. But the scenario was just the reverse during the early history of earth compared to what is found now–life flourishing with rich diversity. In other words, since the earth was born (around 5,000 million years ago) it remained devoid of life for about 1,500 million years. There was no trace of life over the globe at that time.

Having said that, what are the essential requirements for existence of life? They are; air with oxygen in right proportion, water, habitat with food and nutrition along with co-existence and co-operation among different kinds of organism. These basic utilities for living organisms present today in nature were not available on those days. So let us see how the earth is formed – its broad anatomy. The earth is like a very large spheroid ball containing several spherical layers (see Fig. 1, below). At the centre there is a ball of iron nickel alloy – perhaps in molten state and at extremely high temperature and pressure – called **core**. Around it is a **mantle** formed of heavy rocky material in a state of perhaps dense solids for high pressure, or dense liquid because of high temperature. The outermost thin layer of the solid globe is the **crust** made of rocks and soil. This part of earth alone is in our reach and seen directly on surface. The surface of earth is not at all plain, nor uniform. Here there are continents and islands – the land surface including mountains, hills, plateaux and plains – making one third of the globe and the encircling seas and oceans making two thirds.

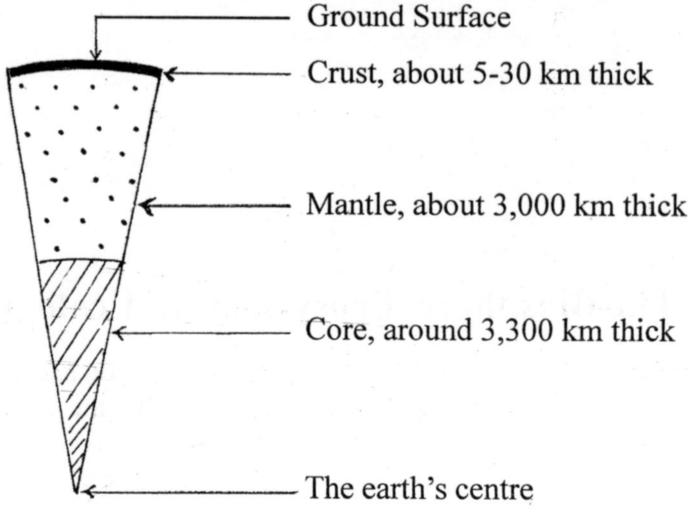

Fig. 1 : Structure of the Earth's Interior

The entire crust (geologically along with the upper hard part of the mantle) – basically made of rocks and soil and having uneven surface–is termed **lithosphere**. All the water bodies of earth (seas and oceans, lakes, rivers and streams), held by lithosphere over its depression, together with groundwater within rock and soil are in toto called **hydrosphere.** There is an envelope of air around lithosphere and hydrosphere and that is known as **atmosphere.** The parts of lithosphere, hydrosphere and atmosphere – especially around their interfaces – where life is present, may be collectively called **biosphere.**

It is very interesting that the span of biosphere generally finds place only around the interfaces of atmosphere, hydrosphere and lithosphere. That means – life is sustained on the ground surface or near surface, in soil or in shallow depths of water including seas and oceans. Because, the essential amenities are made available to biosphere by nature from amongst each of atmosphere, hydrosphere and lithosphere in close co-ordination and interaction – not at all in isolation. Moreover, all the ingredients required for life process could have been exhausted in course of time, had they not been renewed continuously by a good number of cyclic processes of nature. There has been free movements of these vital ingredients and energy across lithosphere, hydrosphere, atmosphere and biosphere through these cycles all along since the birth of our planet. The process is still going on to continue for ever. These earth cycles were absolutely physico-chemical in nature until appearance of life and so they are called **geochemical cycles.** After life came in, some biochemical processes were integrated into the cycles making them what are known as **bio-geochemical cycles.** As a collective result of all such innumerable cycles running uninterruptedly, all the essential life-supporting ingredients like water, soil, oxygen, carbon, nitrogen, phosphorus, sulphur, etc. are being recouped continuously in nature. By this virtue of nature life is sustained on earth. Let us visualize this quadrilateral relationship among the four arms (the four "spheres" – lithosphere, hydrosphere, atmosphere and biosphere) of environment in a schematic drawing given below (Fig. 2).

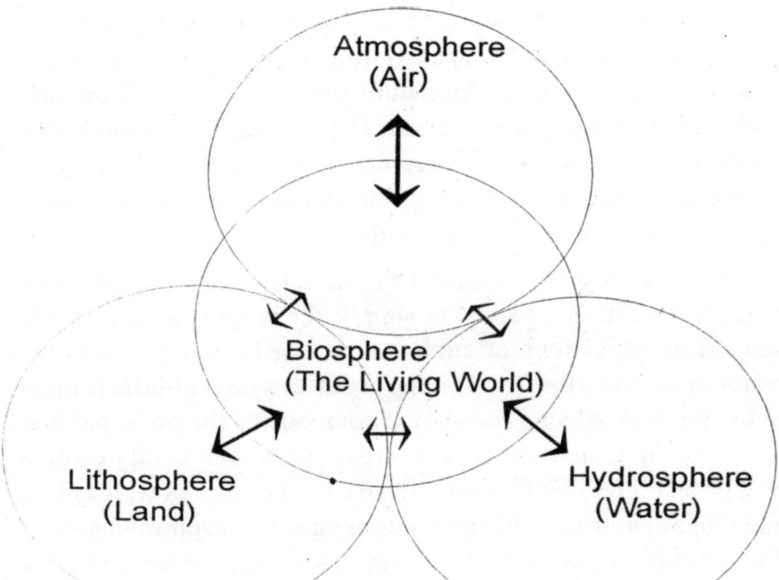

Fig. 2 : The 4-pronged feature of environment

(Note the ambit of biosphere in the zone of interfaces among other "spheres". The arrows indicate movements of matter and energy.)

The biosphere with life spanning over land, water and air has uncountable number of animals plants, algae, fungi, bacteria and viruses belonging to many million species living together in innumerable communities. But the scenario of diversity of life is not the same everywhere. Because the geograhical variation creates different kinds of living enviornment in different parts on the surface of the globe. That is why we find different ensembles of living beings in various geographical settings – in terms of number and species. Somewhere we find dense populations of large number of species such as rain forest. Elsewhere – for example, in deserts or polar regions – very small population of a few species can be found. Deserts, polar regions, mountains, forests, seas, oceans, rivers, lakes, ponds, etc. are geographical entities. Every such geographical entity having unique association of organic species may be called **ecosystem.** In other words, all the ecosystems together constitute the biosphere.

Each of the innumerable species of plants, animals and microbes living together in biosphere get their basic life-supporting elements from air, water, soil or habitat and from other species as well. This system has been working in nature since advent of life on earth without any break or interruption. The science of the relationship between ecosystem and the organisms living in them is known as **Ecology.** It is an integrated part of **Environmental Science** to which I wish to introduce now very briefly.

Every organism obtains all the basic amenities required for life process from its ecosystem, or simply, the environment. The life-supporting provisions of environment can be grouped into two major divisions – **biotic** (of living organisms) and **abiotic** (of non-living bodies). Among the abiotic components the principal ones are : air, water, land/soil/rock, energy, food and mineral nutrition. Soil or land gives habitat for terrestrial organisms as well as food and mineral nutrition. Air provides oxygen for respiration support and, in case of plants or algae, carbon-dioxide for food. It is the medium of pollination as well. Water bodies provide habitat for aquatic organisms. Water – on surface or underground – is the principal medium of metabolic processes with regard to almost all the organisms, be it the case of advanced plants or animals or even

Plate-1

A small virgin island made of coral debris (background) and coral colony (foreground) in Lakshadweep Islands– a shallow marine ecosystem

microbes. These apart, water is the essential raw food material for plants and algae in order to photosynthesize with carbon dioxide. Moreover water is needed for drinking by all higher animals including man while man uses water in every walk of life – at home or outside, in farming or in industry. Energy too is essential for all forms of life. For advanced animals it is obviously needed for locomotion in the mechanical form as it is required for all the metabolic processes of all organisms, in other forms. This energy, chemically stored in the body of all organisms, is released through the process of respiration. Through the reverse process of photosynthesis, energy is captured from the source of sunlight and stored in the bodies of plants. It is, in turn, transferred from plants to animals through food. So sunlight is the ultimate source of energy in the entire biosphere.

Now let us look at the biotic components of environment. An individual organism or an organic species can't live alone on earth. For feeding, reproduction and many more vital activities a living being as well as a living species depend upon another or even more. A large number of organisms living together in close co-operation form a **population.** Similarly few or many populations of different species living together with a relationship of interdependence make a **biotic community.** Interdependence in ecosystem is of various types, such as; 1) producer – consumer relation, 2) predation, 3) scavenging, 4) parasitism, 5) commensalism, 6) symbiosis, 7) competition and other sundry relationship.

Producer – consumer relationship is the most common or nearly universal phenomenon. All plants along with some microbes and algae make their own food by way of photosynthesis. Since they produce organic matter – as food for them and all others – from inorganic substances they are called **producer.** Fishes and some other aquatic animals feed on phytoplanktons like green algae.All plant-eating animals get their food and nutrition from plants while some animals feed on the flesh of plant-eating animals. So all the animals, thus depending upon plants for consumption of what is produced by plants to sustain life, are regarded as **consumers.** This apart, dead plant debris enhances soil quality – for new-generation plants whereas fungi and microbes grow on decomposed plant debris.

A slightly deferent kind of food relation is **predation,** found among carnivorous (flesh – eating) group of animals. Here the predator animals hunt other (prey) animals for their food. Good examples are all carnivores including tiger and lion along with some reptiles (like snakes) and birds among others. Tiger or lion hunt deers, zebra or other innocent animals. Snakes prey upon frogs and toads who in turn feed on insects. In an ecosystem – say, the sundarbans, for example of a forest – all the predators co-habit with their preys making the system viable.

Some animals and birds such as jackals, street dogs, vultures, crows, etc. feed on the flesh of dead animals or waste food. This type of food habit is called **scavenging.**

It is commonly found that some bacteria, viruses, fungi or other animals feed on the body of other live animals or plants while residing there. Thus they get shelter and food from the host organisms causing damage to the latter. This sort of relationship with benefit to only one partner causing damage to the other partner is termed **parasitism.** The invading partner is called **parasite** while the invaded organism is called **host.** Take, for example, mosquitoes living on plants and sucking juices from leaves, flowers and fruits. They even suck blood from animal bodies including human. Similar is the case of lice. Worms live within animal or plant bodies feeding on the latter. Besides, there are innumerable microbes infecting plant or animal bodies causing disease.

Are, then, parasites or predators, who live at the cost of other, undesirable in an ecosystem? Of course not. For every species has its role to play in an ecosystem or in the biosphere at large. A species that causes harm to another may be useful to some other species and, so, is definitely desirable to be in place.

Sometimes it is found – though not so frequently – that an animal of a particular species stays on the body of an animal belonging to some other species giving benefit to the host. It does not harm the host, doesn't take any benefit in exchange. This type of association is called **commensalism.** There are a few instances such as the attachment of sea anemone to the body of hermit crab, barnacles to a whale, or remora fish to a shark.

A very interesting relationship is found in some groups of organisms where there is a mutual type dependence between two species. Here, the two species live closely together without harming either but getting benefitted from each other. This relationship is known as **symbiosis.** Let me briefly explain with a few examples. A bacterium named Rhyzobium lives in nodules grown over the roots of some beans group of plants. Rhyzobium fixes nitrogen from atmosphere which gives nutrition to the host plant whereas the former gets organic food, water and mineral nutrition from the latter. Similar is the case of lichen which is a typical symbiotic association of an alga and a fungus growing upon rock exposure or bark of tree in intimate cohabitation. The fungus absorbs and store water which is used by the green algae to survive and photosynthesize food for growth of both. You may find many more examples such as bees sucking nectar from flowers while helping pollination, and so on.

Contrary to such co-operating and mutually beneficial relationship, there are instances of **competition** as well. As in human society, competition is useful for the biotic community too. It may be amongst the individuals of a species as well as among different species. The significance is clearly displayed in a forest with big trees, smaller trees, shrubs and herbs growing together competitively to get water and mineral nutrients from the soil as well as air and sunlight from the atmosphere. On the other hand, one can easily visualize the competition among predator animals in catching a prey. Competition helps develop and enhance biological fitness and adaptability to environment through generations, as we know from the concepts of "struggle for existence" and "adaptation" mentioned by Charles Darwin in his Theory of evolution.

There are many other sundry cases of inter-dependence in the living world. But further details need not be given here. However, what essentially needs to be mentioned here is the term called **biodiversity.** The term means the diversity of organisms in terms of number of various species as well as abundance of each species living together in a given ecosystem. The more the biodiversity, the better is the effect obtained through interdependence in an ecosystem -- especially for balanced supply of food and energy required to sustain life process. Then how do all the organisms acquire food, nutrition and energy from the environment? Is the practice same for all the species? No, there are various practices for different species. Let us come to that point.

The earliest organisms – some primitive bacteria – could produce their own food chemically from inorganic substance bearing sulphur, iron, etc. They are termed **chemo-autotrophs** for such feeding practice. Later in course of evolution green algae and plants came in. They were able to synthesize – photochemically with sunlight – organic food called carbohydrates with carbon dioxide from air and water from soil or water bodies. This was possible by virtue of green chlorophyl contained in them. This process is called **photosynthesis,** and such organisms are termed **photo-autotrophs.** Energy is chemically stored in carbohydrate to be liberated again though respiration. Chemo-autotrophs and photo-autotrophs being able to make their food directly from physical environment are together called **producers** (of organic matter). On the other hand, there are some microbes, fungi and all the animals who are unable to synthesize organic food directly from the physical environment. They have to depend upon plants or other organisms for food and nutrition and so are regarded as **consumers.** The group of animal feeding solely on plants is known as **herbivores** and the group that feeds solely on flesh of other animals is called **carnivores**. The animals who eat both plants as well as animal flesh are known as **omnivores.** Thus there are three types of food habit among animals – plants or vegetable matter, animals or flesh and mixed.

This is how food or organic matter is produced and energy captured in the biosphere from physical components of environment by the producer group of organisms to be consumed by themselves along with the consumer group of organism for growth, nutrition and energy. As herbivores directly consume plants/algae (the producer), for these basic causes, they are called **primary** (first level) **consumers.** Since carnivores and omnivores consume primary consumers to get food, nutrition and energy, they are called **secondary** (second level) **consumers.** Similarly, there may be third or further levels and there are other organisms like fungi and bacteria who grow on dead and decaying bodies of plant or animals. They are termed **decomposers.**

Anyway, the above is, in short, how variously food, energy and nutrition are harvested by different organisms to subsist in the biosphere. In this process, a chain-type food relationship is established among various species in each ecosystem and the relationship is termed **food chain.** Take this example to illustrate the case. Various insects feed on leaves, flowers and fruits of plants available in a forest ecosystem. Frogs and toads subsist there with preying upon insects. There are snakes, in the same forest, who gulp frogs and toads (among other animals) for food. There may even be peacocks and peahens who eat the flesh of snakes. On death, any living body is decomposed and eaten up by microbes. In this way, a chain is made with the successive links as I said. It looks like as follows : plants → insects → frogs & toads → snakes

→ peacocks & peasans → microbes (decomposer). Similarly another chain may be, say in Sundarbans, like this : plants → deers → tigers. In an aquatic ecosystem, say, in a sea, a big estuary or a lake, there are phytoplanktons (tiny green organisms floating on water such as algae), thriving and photosynthesizing, only to be eaten up by fishes or some insects. Some big fishes eat smaller fishes or insects along with phytoplanktons. The fishes, in turn, are preyed upon by crocodiles or alligators living in that water body, if any. Thus we get the chain like this :

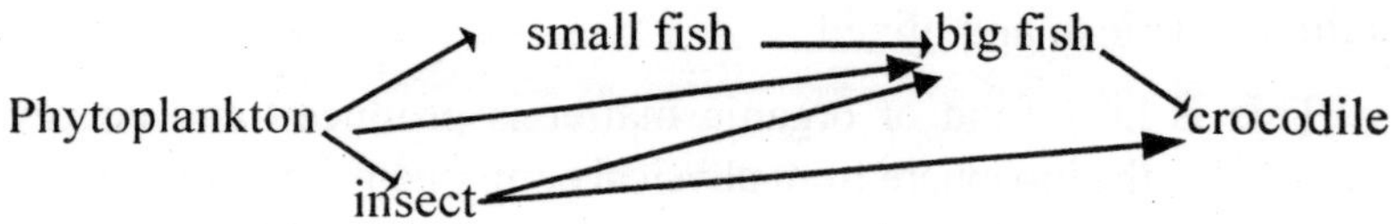

Or, look at another case of a forest:

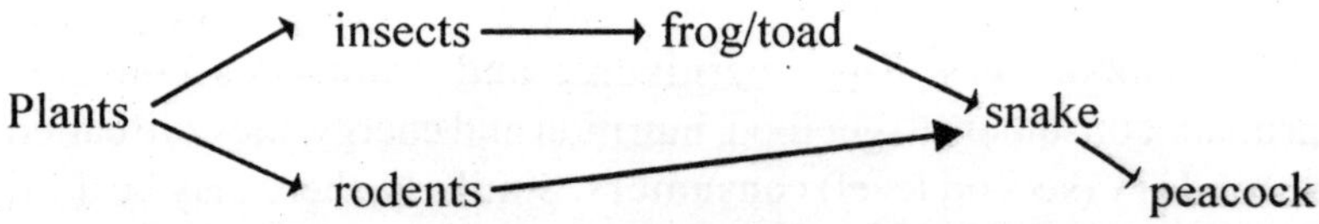

Do you find any difference from what we saw in earlier chains? Yes, instead of simple linear chains there are branching and merging links forming loops and making the picture like net or web. This kind of complete food relationship is known as **food web.** There are too many food chains and food webs in the biosphere.

So we see, organic product and energy are captured and stored in the biosphere step by step through food chains and food webs. But both organic matter and energy are considerably lost while climbing up each tier of the chain. The loss is around 90%. That means only 10% of the organic product and energy of a given trophic level can be consumed and transferred to the next tier up the food chain. It is assessed in terms of size of the population. Thus, in a forest, population of toads and frogs would be around 10% of insect population and snakes would be 10% of toad and frogs, if there be a single food chain – thereby population size being gradually reduced up the tiers in food chain. If this is graphically viewed it looks like a pyramid, and so, this scenario in an ecosystem is called **pyramid of number and energy.** One can see, for example, gradual reduction of population size of the plants, deers and tigers in a forest ecosystem. I have attempted to illustrate the pyramids for a forest and an aquatic ecosystem with the drawings given below.

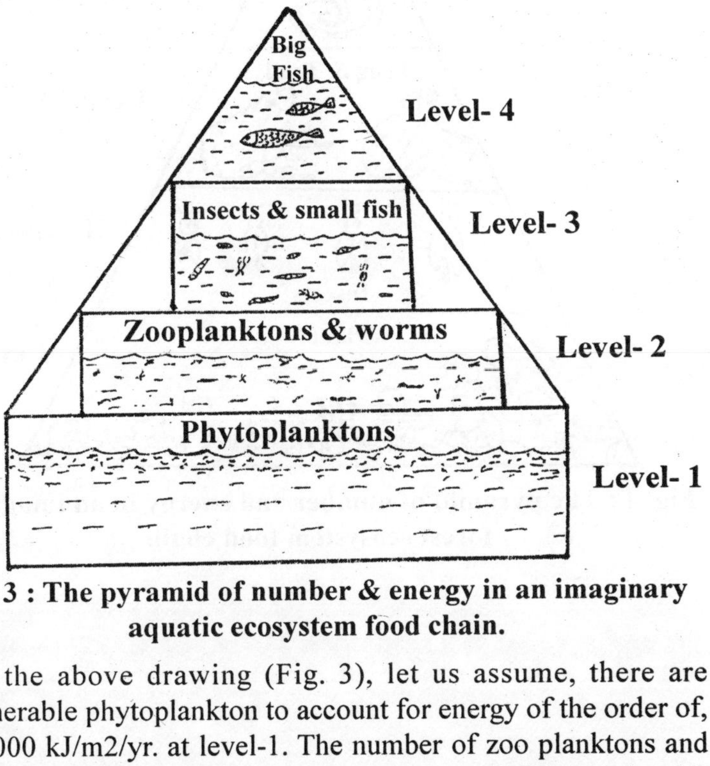

Fig. 3 : The pyramid of number & energy in an imaginary aquatic ecosystem food chain.

In the above drawing (Fig. 3), let us assume, there are innumerable phytoplankton to account for energy of the order of, say 5000 kJ/m2/yr. at level-1. The number of zoo planktons and worms at level-2 together with insect and small fishes at level-3 subsisting on phytoplanktons, the energy is reduced to around 500 kJ/m2/yr. and the numbers around, say, 1000. At level-4 the number of big fishes is expected around 100 and energy about around 50 kJ/m2/yr. Similarly for the forest ecosystem (Fig. 4) against the plant biomass accounting for, say, around 5000 kJ/m2/yr. energy (level-1) the insects present would be of about 500 kJ/m2/yr. (level-2) energy equivalent and, say, 500 in number. At level-3 there would be about 50 frogs and toads (50 kJ/m2/yr.) and at level-4 only 5 snakes (5 kJ/m2/yr). This is however a very rough approximation.

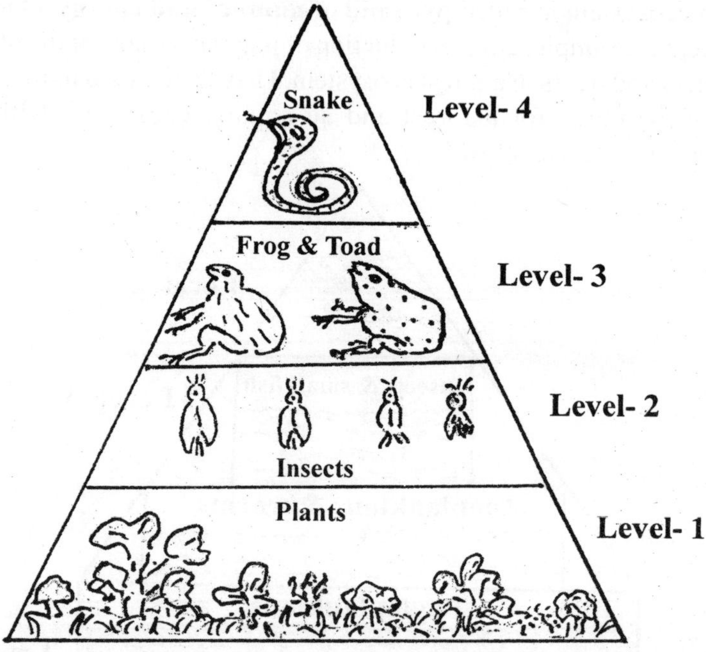

Fig. 4 : The pyramid of number and energy in an imaginary forest ecosystem food chain.

2

Human Society–Evolution & Development

So we have come to know about the wonderful diversity of life – in terms of number of species and variation among them – called biodiversity, present on earth. But the picture was far different around the time the earth was born, about 5000 million year ago. The earth was absolutely devoid of life then and remained so for about 1500 million years as the environment was not conducive of life. This fact, albeit amazing, is a hard truth evidenced by fossil record preserved in the earth's strata. The oldest known fossil record is 3500 million years old. But gradually thc cnvironment evolved into a life-supporting one through its chemical evolution leading to advent of life. Thereafter happened biological evolution at last giving rise to appearance of mankind in the living world called biosphere. Let me tell the story, in short, how it happened.

There was no oxygen in the atmosphere of the young earth. In such a reducing environment, somehow, a sort of broth of some complex organic compounds was synthesized naturally with various elements and simple inorganic compounds present in atmosphere, hydrosphere and lithosphere at their common interfaces, i.e., shallow seas or lagoons. This event is known as **chemical evolution.** The broth had self-duplication DNA molecules. In course of time DNA molecules evolved into living cells encased in membrane. But those unicellular anaerobic organisms didn't have a distinct nucleus. This group is termed by scientists as **Prokaryota.** This form of life, in course of time, gave rise to an advanced group of organisms – at first unicellular but afterwards evolved into multicellular form – called **Eukaryota** having well – defined nucleus. This happened through what is known as **biological evolution**. Thus appeared the plant kingdom along with green algae having green chlorophyl in their body. Chlorophyl enabled them to produce carbohydrate food through photosynthesis out of carbon dioxide from air and water from water bodies or soil. At first there were unicellular or tiny multicellular algae in water and later diverse kinds of plants of various form, structure and size came in. The plant kingdom comprised herbs, shrubs and trees on land and in water as well. This event brought about a revolutionary change in the earth's environment resulting in appearance of molecular oxygen, so vital for most organisms. A delicate balance was then established between production of oxygen by photosynthesis and its consumption by all aerobic forms of life for respiration.

Much later in the history of life on earth – at recent past, to be more precise, and geologically at the onset of Quaternary Period (around two million years ago) – man appeared on earth. In the preceding Tertiary Period, the earth was ruled by mammals (mainly on land) from which man, the **Homo sapiens** has evolved. The first era of life on earth, termed Paleozoic Era, was that of fishes and insects thriving in seas followed by Mesozoic Era being that of reptiles, including the enigmatic dinosaurs, reigning on land as well as in seas. Mesozoic was the era of terrestrial plants, as well, making dense forest all over the land. Anyway, in the first half of Quaternary Period – called Pleistocene Age when the reptiles had already made room for mammals to rule the organic world – apes or the early man appeared on earth. At this time occurred global glaciation which caused extinction of a large number of mammal species. This is evident in the fossil record preserved in river-laid strata of Siwalik hills belonging to the foot-hill ranges of the Himalayas.

However, at the close of this glacial age around one million years ago, when the earth's surface became a bit warmer, there appeared true man (a mammalian animal). This species has been scientifically named **Homo sapiens** (Homo = man, sapien - wise) or the wise man. That is to say, this animal species has the intelligence which is not found in other animal groups, not even in apes. Man is now the ruling animal in the biosphere by virtue of its intelligence and environmental adaptability. The time segment in the history of the earth marking the presence of mankind is, but, very small. It is of just one million years, scientifically called Holocene (recent) Age.

The animal group just evolved as man dwelt in forests and caves. They used to subsist on various types of forest produce like leaves, roots, fruits and meat of preyed animals or birds. They had omnivorous food habit but were unable to cook. They didn't know to build homes, neither farming, nor animal breeding. They were able only to cohabit with wildlife. Forests – albeit everywhere at that time – happened to be their ecosystem. Even then, the level of intelligence and physical fitness of man were much higher than that of any other animal, though the physical size and strength were much lower compared to many other animals. Thus it could quickly establish itself as the ruling group through competition among different animal groups.

While wandering through ubiquitous forests man someday discovered that new tree had been born out of seeds in fruits fallen from trees on ground or from seeds thrown away after eating fruits. In this way they learnt to sow seeds and grow useful plants. Later, they further found that plants grew better on plains – especially beside rivers where soil were better and water-bearing–rather than over hills, plateau or deserts.

It was also realized that it could be more advantageous to grow useful plants at a place and harvest vegetable food-stuff than to wander in forests for the same. Equally advantageous was found to breed animals and get meat than to hunt animals – a very tough job – all the time. Human beings thus learnt to settle at a place making clusters of huts and they got used to farming and animal breeding at that particular place. A village sort of thing evolved in this way. While farming and breeding animals they found that plains along the river-banks or sea-coast were the best sites for the jobs. So human habitations grew more and more over there where man could get adequate supply of basic amenities for living. The amenities are : good soil-bearing land for better farming for crops and animal breeding, source of water for the same as well as for domestic use, level ground and good earth for hut-making, and so on. On being more civilized in course of time, mankind learnt to wear and cook and so were able to gather material to wear and make utensils to cook with. By that time human populations had realized the advantages of living together in communities with close co-operation in an organized manner leading to the evolution of **human society.**

With growing population and rising needs keeping pace with civilization of human society mankind required more and more land for settling habitations, develop housing, agriculture, animal and fowl farming, roads, markets, etc. For these, people cleared forests and established villages to live in society and developed land for farming and cattle-fowl farming in large scale. With further societal development, they started trade and commerce in addition to farming and cattle – fowl breeding. That required roads and waterways for transport and markets for trading. Plains alongside river or sea were again found suitable for these activities as well. The villages gradually turned into growing towns, and the markets, grew larger and larger, over erstwhile forest land, along river-banks or coastlines. This is how the ancient civilizations such as Indus, Egypt, China, Babylon, etc. – all on plains along river-banks or coastlines – were developed.

So the course of evolution of human society started from agriculture and cattle-fowl breeding and passed through organized living, then primitive trade with exchange of farm produce as well as cattle & fowl, then cottage industry to, finally, advanced trade and communication facilities. Later in course of commercial development and scientific inventions mills and industries came in through what was known as Industrial Revolution of Europe in the nineteenth century. Side by side, social and economic development of mankind took place leading to evolution of modern life style. Towns and cities appeared in place of villages. Need for modern homes with furniture and fittings, processed food, purified water, fossil fuels, fancy dress, cosmetics and many more things was felt in place of basic food, clothes and huts in vogue during primitive days. And the same was achieved. We further needed education, healthcare, transportation, communication, entertainment and other services leading to development of schools and college & universities, hospitals & nursing homes, diagnostic laboratories, roads & waterways, railways, shipping, cars, airlines, electricity & appliances, mail and telecommunications, banks, shops & markets, hotels & restaurants, theatres & movies, so on and so forth. Above all, however, there was development of industries (factories, mines, plants and workshops) through which we got all the commodities in markets like paper, books, utensils, implements & equipments, vehicles, medicines, cosmetics, fertilizers and many more consumer products. Industrial Revolution occurred in Europe during nineteenth century and spread over the world thereafter. Green Revolution took place in India during nineteen sixties – seventies through introduction of modern agriculture. This was followed by recent development of high technology in computer and telecommunication. This is the path of socio-economic development of human society that ushered in substantial uplift in the quality of life.

I would like to mention one thing here invoking the 'competition' concept of the theory of biological evolution. Despite the tremendous development achieved in human society (as stated above) all the sections of society couldn't uniformly improve their quality of life. Different races belonging to the biological species **Homo sapiens** evolved at different places over the globe such as Aryans, Dravidians, Mongolians, Caucasian, Negro, Red Indians, Santhalas, etc. They have socio-economically developed in varying degrees and varied ways. It is actually found that tribals have been lagging far behind non-tribals to improve their standard of life. They have been holding back to forests or fringes thereof adhering to forest-based primitive life style. They are really far behind in the present race of development.

3

Man & Environment– Utilization of Natural Resources

So far we have mostly looked at mankind as an animal species **(Homo sapiens)** – its natural identity as a matter of fact – and its relation with nature in the light of biological or ecological science. We shall now view in the light of sociology and economics. Because, unlike common creatures, human needs are not limited only to subsistence and reproduction. Modern people need much more – healthcare, education, comfort, entertainment, tours, transport, communication, trade & commerce, property, money, etc. And all of these are basically obtained from nature – the surrounding around us. This surrounding of ours wherefrom we fulfill our basic biological, social and economic needs together make what is known as **"environment".** The science dealing with environment and its relationship – biological, social & economic – with mankind is termed **environmental science**. It is important to distinguish the latter from ecology though these two are closely allied and we often allude to each other.

We have earlier seen the four-pronged identity of ecological environment consisting of air, water, land and biota along with the energy contained in them. This definition of environment is valid for all creatures barring man as sustaining life and reproduction is the only concern for them. But biological evolution has raised mankind to such a level where needs are manifold and much more, as told earlier, in a system called society. Man's environment is, therefore, five – pronged, society being the fifth arm. Thus man meets all the socio-economic needs, in addition to biological necessities, through what may be termed as **resources**.

A brief discussion about resources is required now. There are so many resources utilized by man for development that we need a classification first. Good number of resources are obviously available in nature whereas many more are created by man as well. But for all those man-made ones natural resources are the ultimate sources. There is no doubt about this. Anyway, as the resources are so great in number and so varied in character that the classification is a very difficult task. Yet let us try it for a comprehensive understanding.

I would like to make the first partition between **material** and **energy** – the two basic forms of resources – in addition to the obvious grouping into **natural** and **man-made** divisions just mentioned.

Energy resource may be renewable which means it is time and again created in nature so that it is not likely to be exhausted. That way, this kind of energy may be called **infinite resource** as well. This group comprises solar energy, biochemical energy from biomass, wind power, hydro-power (energy from running water or waves), geothermal energy, hydrogen, etc. I would remind here that **solar energy** is the ultimate source of all kinds of energy on earth even as we don't have the know-how for its direct utilization in a mass scale. But it is heartening to know that some introductory applications have already begun, in recent years, in powering vehicles, electronic gadgets, lighting, domestic heating, etc., though the progress is rather very slow. Solar energy is captured by plants and algae – through photosynthesis – as **biochemical energy** which in turn is transferred to animal bodies through food chain. The biochemical energy stored in the **bio-mass** (plant and animal bodies) may be released and utilized as heat energy by man. Use of **wind-energy** mechanically to run wind mills is known historically. Present day wind turbines produce electricity considerably in coastal areas, islands, deserts or hill-tops where plenty of strong wind is available. Similar is the case of running water specially in streams and rivers on hilly terrain having strong current. Using the vast mechanical energy of strong current hydro-electricity is generated with turbines at dams across rivers and streams. Use of this **hydro-power** is now a commonplace and it forms a significant share in the energy sector. A much less conventional energy is **geothermal energy** – the enormous heat naturally occurring deep inside the earth's crust. The deeper we go, the hotter is it there. But then, it is not that easy to tap even a little bit of this vast resource. Therefore this huge storehouse of renewable energy – apparently so promising – has remained untapped for want of knowhow. Another promising non-conventional source of energy is hydrogen. Elemental hydrogen, a constituent of water, is highly inflammable releasing plenty of heat energy when burnt with oxygen in a suitable environment. But here again the bottle neck is the lack of cost-effective technology and, so, the status today is just at experimental stage. Anyway, all these sources of energy are all the time renewed in nature through bio-geochemical cycles and that is why these kinds of energy resources have been classified under renewable energy.

Since ancient time mankind has been producing heat energy by burning dry wood, leaves, straw and other forms of bio-mass. Vegetable oils have been used for the purpose afterwards. Now there are coal & lignite and petroleum & natural gas – together called **fossil fuels** – to produce heat. Apart from direct use of heat energy by burning fossil fuels in vehicles, machines, furnaces, etc., the same is burnt in power plant to produce electricity as well. The electrical energy so produced is called **thermal power** which accounts for most of the energy produced all over the world. Coal, lignite (brown coal), petroleum and natural gas are the most commonly available and conventionally used fossil fuels. Much less common and rather non-conventional fossil fuels are coal – bed methane (CBM), oil shale, tar sand, gas hydrates, shale gas, etc. Search for these non-conventional fossil fuels has, however, gained momentum recently, mainly in developed countries, in the face of dwindling oil and gas resources. Even coal resource is not foreseen to last for long. Well, an emerging alternative solution to the crisis of fossil fuels is seen through **bio-fuels** such as bio-diesel and ethanol. They may respectively substitute for diesel oil and petrol as a whole or mixed in some given proportion. But bio-fuels should be easily available in adequate quantity in a cost-effective manner in order to be used extensively. In this context current attention is more and more focused on another non-conventional mineral fuel. That is **nuclear fuel,** bearing radio-active heavy elements like uranium, plutonium, thorium, etc., capable of generating electricity through modern sophisticated technology. The power so obtained is called **nuclear energy**. Its demand is growing day by day although it has the risk of radio-active hazards like the ones accidentally caused at Chernobyl in the then USSR and tsunami-induced at Fukusima, Japan. But none of these three forms of energy, i.e., fossil fuel, thermal power and nuclear power cannot be renewed, once consumed. They are liable to be exhausted in future. For, it took many million years to form coal deposits and oil-gas reservoirs in the earth's crust in an environment free from human interference. It is just impossible to imagine such an environment on earth now and such a long duration for the fossil fuel reserves to last. On the other hand, the radio-active decay of nuclear fuels is an irreversible process and formation of radio-active elements afresh on earth is uncertain even in millions of years. We may, therefore, call all of these three resources of energy as non-renewable. You may look at the chart (No. 1) given to visualize the scheme of classification of energy resources.

Now look at the topic of **material resource** that may be either **natural** (naturally occurring) or **anthropogenic** (man-made). These are the two broad divisions in the first place. The **natural material resources** too, as in case of energy, may be of two types – renewable and non-renewable. Furthermore, natural renewable substance may be **organic**, i.e., derived from the bodies of organisms. It may be, otherwise, **inorganic** as well. In the organic group, there are forest & forest produce, fauna or wildlife, algae & algal material, fungi and microbes of human use. The importance of flora and fauna as useful resources is unlimited – not only for biological purpose like food but also for many economic reasons. This apart, the role of every organism in the biodiversity and maintenance of ecological balance is simply undeniable. The treasure houses of this kind of resources (flora and fauna) are chiefly forests, water bodies, like rivers, streams, lakes, seas and oceans. These water bodies are the treasure-houses of algae and corals as well as fishes, shells and many more important organic resources. Innumerable microbe species occur almost everywhere – on land, in water and in air.

are basically of three genetic groups – igneous, sedimentary and metamorphic. The most primitive of them is the group of igneous rock types originated from magma, i.e., molten rock-material occurring deep inside the earth. Magma moves up to surface through craters of volcano or through deep fissures and flows over the ground to form, on cooling, layers of solidified lava flows. Or, otherwise, it may slowly uprise through weak zones or fracture voids of strata to occupy any available space intrusively in the earth's crust and then slowly solidify into various shapes of igneous rock bodies. The other two groups of rock are derived from igneous rocks. However, all types of rock, on being exposed, are decomposed, fragmented and partially dissolved by the physical and chemical actions of air, water, ice and (after appearance of life on earth) organisms. The rock debris so formed is afterwards transported to low-lying basins of rivers, lakes or seas to be deposited there physically, chemically or biochemically to form layers of sediment. The latter, being deeply buried over a very long time, is compacted, cemented and thus lithified into sedimentary rocks. Both the rock groups–igneous and sedimentary–at great depths, especially under the impacts of extreme heat of any magma body or any mechanical stress due to crustal movement or deformation, are transformed into entirely different types of rock. Such new rock types with drastic change of structure, texture and mineral composition are termed metamorphic ricks – the third group. All three groups of rocks, being exposed on surface, undergo weathering, fragmentation, erosion and sedimentation again to form sedimentary rock. Sedimentary rock may also be transformed into metamorphic rock when buried at great depth and subjected to crustal stress or contact with large igneous rock-body. All kinds of rock may be molten into magma to regenerate igneous rock.

Let me further explain, in short. According to the modern theory of plate tectonics, the lithosphere consist of a number of rigid rocky plates joined together as may be seen in a football. There are three types of stress at three kinds of plate joints. Stress at some joints is compressive, at some tensile and at others it is shear stress. At compressive plate joints rock strata bend down to great depths resulting in generation of magma by partial melting. The latter moves up along fissures and solidify over surface or near surface into igneous rock. On the other hand, lava flows erupt and spread across tensile plate joints on sea floor to create layers of igneous rock over mid-oceanic ridges.

So what did we see? We saw magma (molten rock) leading to igneous rock, that in turn to soil and sediment which lithified into sedimentary rock. Both igneous and sedimentary rocks transformed to metamorphic rock which later partially melted into magma to move up and form igneous rock again. And this process has actually repeated time and again, during the long history of the earth, in cyclic form called rock cycles. Please see the schematic illustration (Fig. 6) to understand more comprehensively.

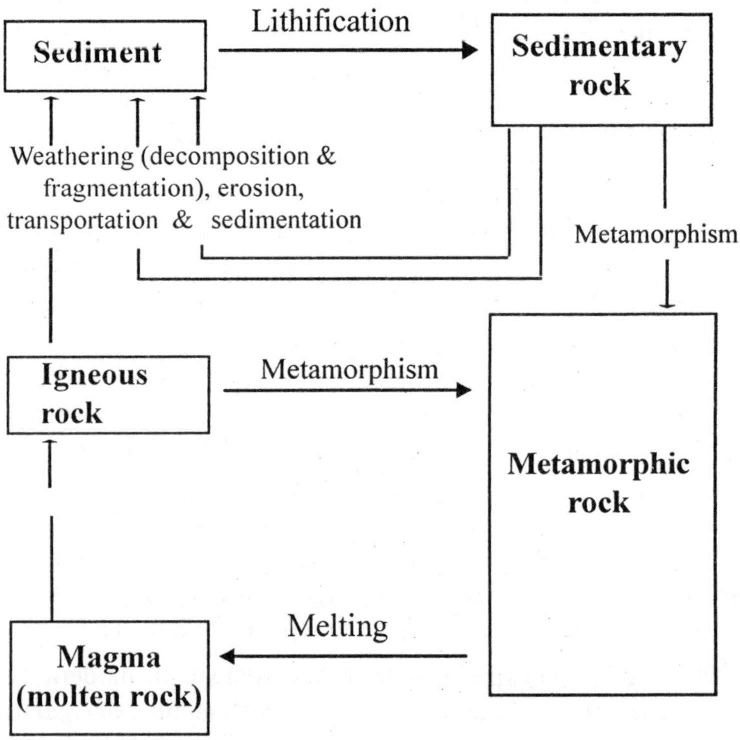

Fig. – 6 : The Rock Cycle (Schematic Illustration)

Now I move on to the **carbon cycle,** another very significant one operating in the biosphere. Air contains 0.03% carbon dioxide. Carbon is the essential constituent of living bodies as well as of any organic substance. We know, carbon dioxide of air and water from soil or water bodies are photosynthesized into carbohydrates with the help of chlorophyl occurring in plant leaves or green algae. Carbohydrates are the basic components of plant bodies and they supply food and nutrition to all the herbivorous, carnivorous, omnivorous and decomposer organisms through food chain. On the other hand, all the energy-giving constituents of living bodies like carbohydrates, fats and oils get oxidized all the time with oxygen of air through respiration process. This leads to liberation of carbon dioxide (along with water vapour) once again into air. The remaining organic substance after death (bio-mass) liberates carbon dioxide through putrefaction, oxidation and other decay process. In this way the flow of carbon from atmosphere to biosphere and therefrom back to atmosphere takes place making a cycle which is repeated all the time. This cyclic process is known as carbon cycle (schematically shown in Fig. 7). Some other events uncommon in nature, may be added to this cycle. But they play very minor roles. Anyway, as a matter of fact the carbon-dioxide content in air remains stable by virtue of the carbon cycle. But various human activities like agriculture, industry and transportation are disturbing the stability by enhancing the carbon content in air. Methane gas is generated through farming and carbon dioxide emanates from industries, transportation on surface, air planes or ships, power generation, etc. We shall come back to this point later.

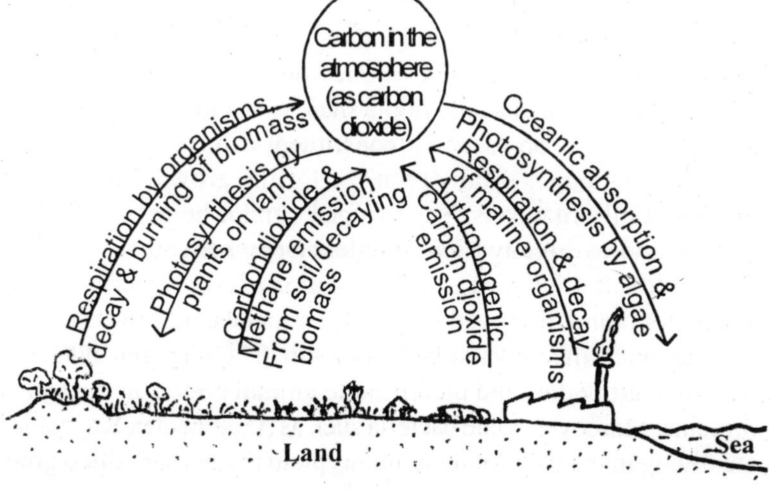

Fig. 7 : The Carbon Cycle (The arrows indicate the flow of carbon)

Now look at another very significant cycle – the oxygen cycle – so vital for the biosphere. Oxygen as gaseous element is present in air in a fair proportion (about 21%). Moreover, the enormous volume of water in the hydrosphere has oxygen chemically combined with hydrogen. It is present in all the biomass, as well (the quantity albeit being small), combined with carbon and hydrogen. No living being – be it flora, fauna, algae, fungi or microbes – cannot exist in an environment without free oxygen, barring a few anaerobic bacterial species who can live without oxygen. For, the organisms need oxygen for respiration, the very essential vital process. Most of the aquatic organisms including fishes, crustaceans and algae among others take dissolved oxygen from water for respiration – not to mention all the terrestrial lives inhaling it directly from air. Oxygen is dissolved into water from air through any kind of agitation like current or waves. Is oxygen thus getting depleted in the atmosphere? Is it going to be exhausted some day? No, because oxygen is recycled into atmosphere through photosynthesis by plants and green algae. This is how the oxygen cycle takes place in the present-day nature. (I say "present-day" in the sense that the oxygen cycle had been of different nature at various phases of the long history of the earth with changing biodiversity starting from a life-less era). Apart from the above general profile, some natural processes of oceans and lithosphere, e.g., volcanic activity, also contributes to the oxygen cycle.

Lastly, the **nitrogen cycle.** This too is very important so far as organic growth and nutrition is concerned. Nitrogen in the state of gaseous element makes the lion's share (around 78%) in air of the atmosphere. It is an essential constituent of proteins, a group of organic compounds vitally required for the growth and nutrition of almost all organisms. Yet nitrogen cannot be assimilated into organic bodies directly from air unless it reaches soil or water body as soluble salt. Lightning causes oxidation of atmospheric nitrogen. The latter being dissolved in rain water make acids and ultimately nitrogenous salts in water bodies or soil. The nitrogen salts make way to plant bodies and therefrom to animal bodies in the form of proteins. Moreover, some bacteria such as Azotobacter, Rhyzobium, etc. living in the roots of leguminous plant (beans and allied group) absorb atmospheric nitrogen biochemically to be stored there, or partly consumed by the plants, in usable form. The fungus Micorhyza living near the roots of some plants plays the same role. On the other hand, huge quantities of blue – green algae or cyanobacteria in the oceans fix nitrogen into their bodies to be passed on later to marine animals or simply to water. In this way nitrogen is fixed into lithosphere, hydrosphere and biosphere from atmosphere. The phenomenon is called **nitrogen fixation**. Interestingly, mankind, for their own sake, performs this job as well. How? It is done by synthesis if nitrogenous fertilizers and generation of NOx (oxides of nitrogen) in industries and vehicles by way of oxidation of nitrogen through high-temperature combustion. Anyway, nitrogen fixation goes on in nature all the time. But at the same time, proteins in plant and animal bodies are decomposed through metabolic processes, or through decay after death, into urea and ammonia which are excreted back to soil or water. Therefrom nitrogen gets back to atmosphere. Urea – ammonia may, however, make a short circuit from soil to plant and then to animal bodies through nitrogenous salts and then proteins. Conversely, nitrogenous salts – fixed from atmosphere or produced out of organic excretion or decay–are decomposed into urea-ammonia and then elemental nitrogen by some bacteria occurring in soil or water. This process is termed **denitrification.** Thus nitrogen fixation and the reverse process of denitrification together make the nitrogen cycle.

There are many more such bio-geochemical cycles running for ever in nature like phosphorus, sulphur, calcium, to cite a few. These cycles are responsible for maintaining a delicate system of balance between demand and supply of all the life-supporting natural resources. But the problem lies in human interference into this system due to the urge of creating economic resources in addition to natural ones. Let me leave it for now, anyway. We shall come back to this point at length afterwards.

4

Environment–Quality & Pollution

Through our understanding of the concept of environment so far, we have come to know that a delicate ecological balance is always maintained in nature by way of its own rule. As a result, the life process of every organism (including man) is sustained properly in a definite regime of physical, chemical and biological quality of its environment. Maintenance of this definite state of environment is absolutely essential to sustain ecological balance. Given some amount of slippage from this state of balance, a lot of problem may be created in an ecosystem. An environment or ecosystem in such a state is regarded as polluted and the event is called **environmental pollution.**

But how do we identify an environment – whether healthy or polluted, good or bad? We should, therefore, invoke the concept of **environmental quality** and a well-defined standard for that. By environmental quality we mean the quality of all the four arms of environment, i.e., air, water (surface and groundwater), land or soil and the living world (flora & fauna). So we shall now discuss about all the four arms of environment one by one, briefly.

A question arises before that. How do we assess the quality or measure in quantitative terms? For this task, we are required to invoke some sort of parameters that may help us as yardstick in quantifying different attributes of an environment and these parameters are called **environmental quality parameters.** A particular environment is characterized by a given set of quality parameters. In a good environment a suitable ecological balance is kept up so that no harm is caused to the life of any living being (including human) in the ecosystem. We may, then, recognise such a state of balance (identified with some measured parameters) as the **standard** for the environment. Bio-diversity on earth is enormous. It means there is great variation in the ways of life amongst innumerable organic species. But environmental quality standards are laid down keeping in mind their applicability to human beings or the creatures useful to mankind. There may be some controversy or debate in setting such standard in the backdrop of so many factors to be considered. That is why environmental quality standards are laid down at local, national or international levels instead of a single global standard. The standards are identified with a set of parameters.

The quality of environment remains alright or the environment remains healthy for a certain range of values of given quality parameters in most cases. As soon as the value goes beyond this range (crossing lower or upper limit) the environment becomes intolerable. Such limits of quality parameters are termed **tolerance limits** which are usually the upper limits but may be the lower ones in some cases. Sometimes it is defined by both as well.

As already stated, our environment – a natural environment or an ecosystem, to be more precise – exists as a four-pronged entity. I am talking of air (atmosphere), water (hydrosphere), land (lithosphere) and biota (biosphere) with the energy inherent in each of them. These actually constitute an ecosystem. So we should determine the quality of each of them in order to assess the quality of an environment. Assessment of the quality of biotic environment is, however, very complex and too difficult to measure directly though the degree of biodiversity of an ecosystem may be an indicator. But it is not that hard for the other three physical components of environment. The quality of air, water and soil (or land) can be determined through various parameters with the help of simple analytical chemistry and simple physics.

Having said that, let us deliberate a little on environmental quality parameters. Take, at first, the case of **ambient air quality,** i.e., the quality of air around us at a given location. The most common and significant parameters for this are : **1) Suspended Particulate Matter (SPM), 2) Sulphur Dioxide (SO_2), 3) Oxides of Nitrogen (NOx), 4) Carbon Monoxide (CO), 5) Temperature and 6) Noise Level.**

Some amount of dust particles, regardless of whether visible or not, is always present suspended in air. When the amount goes beyond tolerance limit, the air becomes polluted. This kind of air pollutant, termed suspended particulate matter (SPM) is the most ubiquitous as well as the most significant of all. We see dust in air as a result of stormy wind. We also see smoke emitting from vehicle exhaust, cooking stoves, ovens, boilers, furnaces and chimneys or stack of factories, plants or workshops. The smoke contains a lot of SPM mainly as carbon particles which are dispersed in the ambient air. Dust is generated here and there from tilling or earthmoving activities as well. As for industrial activities, the principal generators of SPM are : mines (mainly the open-cast ones), stone quarries and crushers, thermal power plants (especially the coal-based ones), mineral handling and transportation, cement, metallurgical plants (including iron & steel and sponge iron plants, now mushroomed in India). The ash flying out of the stacks of coal-fired power plants is so fine-grained that it remains suspended in air for a long time and fly to a long distance. That is why it is called fly ash. The coarser fraction of coal ash remains as residue in the boiler / furnace and so is called bottom ash. Fly ash is a burning problem today owing to its very nature just told. Obviously, the coarser or heavier fractions of dust of any kind settles down on the ground at a short distance soon while the remaining finer particles travel a long distance and remain suspended almost until settling down with a rain-shower. These dust particles practically form SPM. The finer fraction of SPM having grain size less than 10 microns is capable of entering into nostrils of man and so it is termed Respirable Particulate Matter (RPM). SPM and RPM are both measured as their concentration in air, with the unit μg of dust per m^3 volume of air, with the equipment called High Volume Sampler. As per the current Indian standard, the tolerance limits of SPM are : 500 μg/m^3 for industrial zone, 200 μg/m^3 in residential or rural area and for sensitive locations like hospital , school or court of law it is 100 μg/m^3 only. Well, μg means one thousandth of a milligram. Remember, this parameter is determined as the average of a total air volume sampled continuously for 24 hours or eight hours.

Fossil fuels (coal, mineral oil and natural gas) contain sulphur. So the smoke emissions from vehicles, cooking stoves, thermal power station and various factories contain sulphur dioxide (SO_2) generated through burning of the fuel. SO_2 readily diffuses in air. Decay of organic matter may also give rise to SO_2 emission, naturally. It may emanate from the earth's crust through volcanic eruption as well. Smelting of metals – particularly base metals like copper, lead and zinc having ores as sulphide – produces this gas in a large scale. It is similarly generated in coke oven and oil refineries as well. SO_2 gets mixed in air unless it is recovered as sulphuric acid before emission. It is a significant air pollutant and so its concentration in air is an important parameter of ambient air quality. The tolerance limit fixed for SO_2 as per Indian standard is as follows : 120 $\mu g/m^3$ in industrial area, 80$\mu g/m^3$ for residential or rural area, and 30 $\mu g/m^3$ for sensitive zones. It is to be measured as average of 24 hours' or 8 hours' continuous sampling, just like SPM.

Any kind of combustion at very high temperature brings about generation of nitrogen oxides (NO_x). This happens through oxidation of nitrogen with oxygen, both from air, in engines, furnaces, etc. NO_x, like other air pollutants, is dispersed after emission. This parameter, usually co-generated with SO_2, is measured similarly with the same units.

Carbon monoxide (CO), generated out of incomplete combustion in confined environment, is another toxic air pollutant. Its presence in air even at very low concentration is fatal. For this reason it is regarded as an important parameter of air quality. The tolerance limits have been laid down in our country like this : 5$\mu g/m^3$ in industrial zones, 2$\mu g/m^3$ in residential or rural areas and 1$\mu g/m^3$ at sensitive places. The values are measured as average of 8 hours' continous sampling of air.

Temperature may also be a notable air quality parameter. For the atmosphere is found comfortable within a certain range of temperature. Beyond that, mankind along with flora and fauna become sick. But air temperature varies with time and space according to climate. Moreover the tolerance limits may be different for different organisms. It is, therefore, very difficult to arrive at a common standard for ambient air temperature which is not generally found in environmental practice.

Lastly comes noise level, a very important ambient air quality parameter, much talked about in recent times. It, however, appears to many people as something else. But the point is that noise–the unpleasant form of sound energy–is propagated as vibration in the medium of air. That way, it is of course an air pollutant and so the intensity level of noise in our ambience is to be regarded as an air quality parameter. According to Indian standard the tolerance limit of this parameter (maximum value) has been laid down as: at daytime for industrial zone 75 dB (deci-Bell) and 70 dB at night time for the same, 65 dB at daytime and 55 dB at night time in commercial and mixed areas (with residential and industrial clusters). It is 55 dB at daytime and 45 dB during night time in case of purely residential or rural areas and in silence zone (e.g. at locations of hospital, court, educational institutions) 50 dB during daytime and 40 dB at night. Daytime means 6 AM to 10 PM and night time means 10 PM to 6 AM. It is harmful to the health of man and other creatures if the noise level exceeds the said limits.

The above standards together constitute **National Ambient Air Quality Standard (NAAQS)** of India. As the term indicates it is valid for all over India with regard to ambient air. Ambient air gets polluted by, or its quality depends upon, emission into air from various industrial or other development activities mentioned earlier. So there are different standard laid down for emissions from specific industries or other development activities. Here, we need not go into that at length.

Next we take up water environment. Water on earth exists in two regimes – surface and underground. We shall call them **surface water** and **ground-water** respectively. Surface water occurs in varieties of water bodies such as seas and oceans, rivers and streams, lakes and ponds, canals, wetlands, etc. These, again, may be grouped into two kinds. One in the group of **inland water bodies** and the other is that of **coastal and marine water bodies**–the former having fresh water and the latter saline. Both types of water are useful for the living world. But for mankind fresh water is much more useful even as it is much less available than saline water. Moreover, it is very scarce at many parts of the globe, particularly over deserts and mountains. Anyway, specific qualities of water are suitable for different usese of mankind. Keeping this in mind, several classes of water quality standard have been formulated in our country for both inland (fresh water) and coastal-marine (saline) water bodies, identified in accordance with best use.

For inland water bodies, they are : Classes A, B, C, D and E. All the primary sources of drinking water are in the ambit of **Class A. Class B** is meant for community outdoor bathing and **C** groups the sources of domestic water after conventional treatment. **Class D** comprises water bodies for fish farming and wildlife conservation while **Class E** is meant for irrigation and industrial use. Since formal deliberation on environmental science at length is not the objective here and only the water bodies belonging to Class A and B are meant for daily use of the public at large, I shall briefly describe here the standards and quality parameters of these two Classes only. But that will come a bit afterwards.

We now come to the saline water bodies in seas or along coast. Here too on the basis of best use they are of five **Classes – I, II, III, IV** and **V. Class I** incorporates salt pan, coastal ponds for farming fish, prawn and allied creatures together with all the coastal ecologically sensitive zones. **Class II** includes beaches for sea-bathing, swimming and water sports as well as the commercial fishing zone of any sea. Sources of the water for industrial use and locations of boating and sight-seeing are under **Class III. Class IV** comprises ports and harbour area while **Class V** consists of the far areas of sea for navigations and limited waste disposal. For saline water bodies too, I think only the first two, i.e., Classes I and II are relevant, in line with inland fresh water bodies. We shall go further on this point briefly, but a little later. Before that we need to be apprised of the quality parameters in general.

The list of parameter is pretty long. I shall cite only the commonly used and generally significant ones. Take **colour** at first. Pure natural water is colourless. If the water is found coloured – be it surface or groundwater – it should be regarded as polluted or contaminated. Next come **odour** and **taste** which vary among waters of different environment. But pollution-free water should be almost odour-less and taste-less. Another important parameter is **turbidity** which indicates how much solid particles are occurring in the water in colloidal form or in fine suspension to make it turbid. The absolute quantity of suspended solid particles occurring in a water sample of unit volume is termed **total suspended solids (TSS).** The unit is mg/l (suspended solid in milligram per litre of water). Measurement of TSS is very simple. All we need is to pass a given quantity of water sample through filter paper and compute the concentration of filtered solid after drying and weighing. Measurement of turbidity in absolute term is, but, not possible. It is assessed in a relative scale by visually matching with a few standard samples. The scale comprises a few numbers like 1, 2, 3 and so on. Such type measuring units of turbidity is called **nephelometric turbidity unit (NTU).** Another very important parameter is **oil & grease** which may be both mineral or organic. Oil and grease float on water. To determine this parameter (i.e. degree of presence of oil and grease), water sample is to be kept in a container or chamber to allow TSS settle down and then the oily surface layer of the water is collected. By recovering the oil and grease from that collection we may compute the parameter. The unit of measurement is here again mg/l (oil in milligram per litre water). Anyway, all of the parameters so far cited are physical ones. We shall move on now to the chemical parameters, but only the principal ones in short.

Let **pH** come at first. It is an indicator of acidity/alkalinity of water – indicating, actually, the reverse of hydrogen ion concentration – expressed as a number. It doesn't have any unit. The value 7 is for neutral water. More than that indicates alkaline water and less acidic.

Dissolved Oxygen (DO) is another important parameter. This is just concentration of oxygen gas dissolved in water measured with the unit mg (of oxygen) / l (one litre of water). Its importance lies in the fact that a minimum amount of DO is required in an aquatic ecosystem in order for subsistence of fish, prawn & shrimps and many other aquatic lives breathing in water. If DO goes down below that value the aquatic life is endangered and so this lower limit is the tolerance limit of dissolved oxygen. Unfortunately DO gets diminished in water bodies owing to the presence of some impurities or contaminants. There may be a lot of organic matter (dissolved or suspended), admixed in water, which are amenable to oxidation by the oxygen dissolved in that water. This oxidation is biochemically facilitated by the microbes present there. Let me explain in short. The microorganisms feed on the organic pollutants voraciously. Then the food is oxidized and destroyed through respiration utilizing the DO present in the water body, for, that is the most ready supply of oxygen. This leads to quick depletion of DO in the water body. The more there is organic pollutants the more and faster is the decrease of DO. So the rate of depletion of DO in a water body may be an indirect indicator of concentration of organic pollutants amenable to biochemical oxidation at a given temperature (since rate of oxidation is controlled with the ambient temperature). The reduction of concentration of dissolved oxygen in a sample of water from a water body determined in a laboratory during a given time in a standard temperature is thus called **biochemical oxygen demand (BOD).** It is expressed as DO loss in milligram per litre of water sample (mg/l) and measured at 20°C temperature continuously for 5 days. But inorganic impurities and some bio-nondegradable organics are not capable of biochemical oxidation. They are of course amenable to chemical oxidation. These too are responsible for loss of dissolved oxygen in water and the degree of their presence can be determined with the loss of DO content similarly as in case of BOD but with a chemical (abiotic) technique. This parameter is termed **chemical oxygen demand** (COD) and measured with the unit mg/l.

Water runs off over soil and rock surface or flows through voids in the latter, to get accumulated in a water body or groundwater reservoir. In this process it takes along some solid chemicals – basically various salts – dissolved in it, however small may be the amount. The total quantity of all the dissolved solids present in a unit volume of water is termed **total dissolved solids (TDS),** reported with the unit mg/l. The filtered liquid after filtering out TSS, as stated above, may be dried up and then the solid residue may be weighed. The concentration of the solid residue per unit volume of water gives the TDS value.

Some of these dissolved salts are of such nature that they inhibit frothing and precipitate on boiling the water to form a crust on the boiling pot or boiler. They are generally bicarbonates and some carbonates and sulphates. This kind of water quality is called hardness and the water known as hard water. Hardness is determined by precipitating the soluble salts and reported as bicarbonate or equivalent thereof with the unit mg/l. Significantly, as I said, groundwater is usually harder than surface water because of dissolving more and more salt while percolating through soil or rock strata.

These apart, there are many more chemical water pollutants. But going into them at length, I think, is not necessary here. I would mention just only one. It is the group of **heavy metals** and other **toxic pollutants**. These are of serious concern as they may lead to far-reaching harmful impact upon human or other living bodies. This group, therefore, is that of very significant water quality parameters. Notable among them are cyanide, fluoride, arsenic, phenols as non-metals and mercury, lead, chromium, cadmium, etc. as heavy metals. With these we may tag radio-active substances like uranium. The unit of measurement is mg/l here again.

Well, in addition to the physical and chemical parameters, there is an important biological parameter of quality or pollution of water. It must be cited before we close this topic. It is known as **coliform** – a group of bacteria causing various enteric diseases. These bacteria sourced from excreta (human or other) is carried and infected through water. The parameter is determined with pathological technique as MPN counts per 100 ml water. I hope this much is quite enough for a common reader to have a fair introduction to water quality parameters. More detailed deliberation – required for formal education – may be skipped here.

Now I get back to the topic of water quality standards for different classes of water bodies based on most suitable use starting with class A (inland direct source of drinking water). The list of chief quality parameters for this is as follows : pH 6.5 – 8.5 (permissible range), DO 6 mg/l (minimum), BOD 2 mg/l (maximum), coliform 50 (maximum, MPN index per 100 ml water sample), colour maximum 10 Hazen units, odour not unpleasant, water without any taste, TSS 500 mg/l (maximum), hardness limited within 300 mg/l, dissolved iron within 0.3 mg/l, sulphate salt 400 mg/l (maximum), nitrates below 20 mg/l, fluoride not exceeding 1.5 mg/l, mercury not exceeding 0.001 mg/l, arsenic below 0.05 mg/l, cyanide not exceeding 0.05 mg/l, mineral oil 0.2 mg/l (maximum). Mind it, this is the standard for any source of drinking water (wherefrom water is fed into filter plant) – not actually drinking water itself. The standard for drinking water, laid down by WHO (World Health Organization) and followed in our country, is much more stringent. But that is not relevant here.

In case of class B (water bodies for community outdoor bathing) the standard is : coliform bacteria not beyond the range 500-2500 MPN index per 100 ml, pH 6.5-8.5, dissolved oxygen at least 5 mg/l and BOD maximum 3 mg/l (for 3 days at 27°C). .

The quality standards for the two relevant classes of marine/ coastal water bodies (as I said only these two generally are of public concern) are as the following. For class I (including salt pan, ponds for farming shrimps, etc., mangrove forest zones like Sundarban and other such ecological sensitive zones) : pH 6.5-8.5, dissolved oxygen (DO) at least 5 mg/l or 60% of the saturation value – whichever is more, no colour and no odour, no harmful suspended solid or dirt, oil & grease maximum 0.1 mg/l, toxics & heavy metals 0.001 mg/l.

Similarly for class II (sea water at beaches chiefly used for sea bath, swimming & water sports and commercial fishing) the standard is like this : pH 6.5-8.5, DO at least 4 mg/l or 50% of the saturation value whichever is greater, BOD not exceeding 3 mg/l (for 3 days at 27°C), no colour and no odour, turbidity not more than 30 NTU, no putrid floating object, sewage generated coliform less than 100 MPN index.

Now we move on to the regime of groundwater. Its quality assessment is similar to that of inland surface water. But in this case, physical parameters like turbidity, TSS, etc. are almost irrelevant while the chemical parameters like pH, hardness, dissolved solids, iron, heavy metals, arsenic, fluoride and other toxic pollutants along with coliform happen to be very much significant. Anyway, since there is no separate standard readily available for groundwater, the same for inland surface water may be applicable for groundwater depending upon its use. Notably, the main use of groundwater happens to be domestic (source of drinking water) and irrigation. Sometimes it is used for industry as well, even as that is not desirable for the sake of environment.

Let us next focus on the scene of land environment. The basic constituents of land are soil, sand, gravel or rocks. Soil is fertile on plains – alluvial floodplains in particular – because of silt alluvium deposited from flood water. Degraded brown laterite soil occurs on uneven plateau. Soil in coastal region is clayey while beaches are sandy and deserts are covered with sand and dust. Besides, there is muck or mud in wet lands and organic humus soil formed of decayed litter in dense forest. Mountains and uplands of plateaux are, however, covered with naked rocks, mostly. Anyway, land is used by mankind principally for habitation and farming. For that we need plain land and fertile soil. Moreover, varying qualities of soil are needed for different crops. Soil quality is defined with several parameters such as grain size, porosity, permeability, bulk density, moisture content, water-holding capacity, organic matter or humus, pH, NPK (the basic mineral nutrient – nitrogen, phosphorus and potash) content and mineral micro-nutrients content. These are the quality parameters of farmland soil, the standard varying depending upon crop to be farmed. But summarily we can say loam (mixture of sand and silt) grain size, good porosity and permeability, moderate bulk density, good moisture content and water-holding capacity, organic matter in fair amount, pH around neutral (7) and good amount of NPK and micro-nutrients make a soil fertile for farming. For other uses of land the parameters may differ. But, on account of large variation of land use, no specific standard is available for quality parameters.

Well, the fact is that land (or soil), water, air – all the three physical environments – are intricately entwined with the biological environment to make the holistic entity of any ecosystem or the entire biosphere. There are innumerable organic species in the biosphere. The diversity of their life style and metabolic process is almost infinite. A definite and objective system of quality standard for any organic environment or ecosystem with biological parameters is therefore very difficult. But there may be some bioindicators in order to know about the health of an ecosystem. For example, you saw that most fishes and shrimps in a pond or any other water body died while other creatures like frog or crabs remained mostly alive. This obviously indicated that DO of that water body had gone down below tolerance limit or BOD risen above. Nevertheless, I think, the biodiversity of a given ecosystem may be considered as a holistic parameter of environmental quality.

As long as the environmental quality standards for air, water, land/soil along with biota (flora & fauna) remain maintained, i.e., all the parameters remain within tolerance limit, the environment is deemed O.K. But as soon as any parameter goes beyond tolerance limit it is deemed polluted. Because it causes adverse impact upon mankind and property along with other creatures. This transformation of environment from a healthy or tolerable state to an unhealthy or intolerable one is known as **environmental pollution**.

So how does it happen? I mean pollution of environment. Sometimes in an environment some of its essential components or resources get depleted or damaged due to various reasons. Otherwise some undesirable or harmful substance or condition may appear in excess of tolerance limit. This results in pollution of the environment. Such substances or conditions are called **pollutants** which are generally identified with quality parameters beyond tolerance limit. Pollution may occur in any of the four arms of environment (air, water, land/soil and biota) or the entire ecosystem. The pollutants may be identified accordingly, e.g., air pollutants, water pollutants, etc. Sometimes there may be a single pollutant, otherwise there may be a number of pollutants together as well. They give rise to impacts accordingly.

Now we shall look into the **pollution processes** which may be of various kinds. It may be simply **natural** such as volcanic eruption. This event suddenly causes emissions of a lot of dust (SPM), heat, toxic gases of fluorine, sulphur, etc. into air. That, in turn, may result in pollution of water and land along with air. Useful land turns into lava flow, heaps of rock debris or dust. Similarly, flood (from heavy rain shower), cyclone, earthquake or tsunami may lead to epidemic or severe damage to land/soil or quality of water in water bodies or groundwater. On the other hand, the pollution of water (groundwater in particular) and soil indicated by arsenic, fluoride, pH, iron, hardness are all geologically caused. The origin lies in different geological processes as well as mineralogical composition of strata. Forest fire is another natural event that may cause air pollution. But by far the most significant illustration of natural pollution (or environmental disaster, to be more precise)–albeit catalysed by human activity–is successive droughts leading to desertification. Besides, unusual cold wave or heat wave many also be worth mention. But such cases of natural environment pollution are not so conspicuously felt as are the impacts of **anthropogenic** (man-made) **pollution.**

So let us now focus on anthropogenic environmental pollution. A notable objective of development in human society is to produce useful resources or products required for value addition to the quality of life. Starting from the primitive activity of agriculture and animal farming through ultra – modern industries and services of high technology, everywhere some useless substance is inevitably generated along with useful products or by-products. Such useless matter is called **waste** which enters into environment in the form of **gaseous emission, discharge** of **liquid effluent** (or waste water) or **solid/semi-solid residue.** These wastes – no matter whether solid/semi-solid, liquid/waste water, gaseous/dust emission – contain various kinds of environmental pollutants to which we have already been introduced. Different pollutants alone or together are liable to cause harmful impacts resulting in damage to the health of man and properties along with other creatures, biodiversity and the overall ecosystem.

We should, then, identify the human activities which lead to environmental pollution and know how it happens. Almost all the developmental activities are the causes of environmental pollution, more or less. The most primitive and by far the largest human activity is **farming** that accounts for large-scale pollution. This is a bare fact even as it sounds strange. There is a thick mantle of fertile soil over plains. So at least 2-3 rounds of crop a year are farmed intensively over there. During dry season the fields are irrigated with rampant use of groundwater through pit wells, tube wells (deep or shallow), ponds. As a result, water table (the top surface of groundwater saturated in the pores of strata) lowers down fast especially in case of deep tube-wells. So it may give rise to a crisis called **groundwater famine** in future. This aside, with the urge to quickly augment crop yield, chemical fertilizers and pesticides are applied too much now-a-days. These two chemicals get washed from the crops or the soil with rain shower and get into the waters of rivers, streams, lakes and ponds. By way of irrigation and groundwater percolation they contaminate groundwater as well. Ultimately through food chain these two harmful chemicals enter into the bodies of man or other creatures to cause many health problems. Generally the impact is not felt at once on account of very small doze. But with continued assimilation in the body these harmful chemicals may accumulate enough to cause serious consequences. This phenomenon is known as **bio-accumulation**. It has grown to an alarming position in recent time. Who knows if this may be the cause of such outbreak of cancer seen today?

So we see, pesticides and fertilizers get into crops and water bodies. Then animals including man feeding on them get contaminated with these chemicals through food chain. And step by step up the food chain their concentration in organic body gets magnified many times the original value present in crops or water. This event is termed **bio-magnification.** Let me demonstrate like this. A pesticide gets into the body of a man through the crops of a farmland. The same also finds way into the bodies of cattle feeding on grass and crop wastes of the field. The same man eats meat and drinks milk of the infected cattle and adds the chemical further in his body. On the other hand, the insecticide goes into the water in a pond alongside the field and then into fishes there. It then gets into the same man's body through fish food and water. In this way the toxic chemical enters into a human body progressively through several routes and year after years. Having said that, what would be the quantum of pesticide accumulated in that man's body after many years?

There is another important story involving fertilizers along with organic manures used in agriculture. Fertilizer means supplementary source of rich nutrients applied to soil in order to enhance crop growth and yield. As fertilizers are water-soluble, there is every chance that they may be rain-washed from crop fields and flow with water into water bodies nearby, if any. Actually this happens often and the water in the said water body becomes highly nutrients-rich resulting in rapid growth of algae and water hyacinth. When this growth reaches the maximum, it leads to their sudden death en masse. Then the natural decay of the dead mass by oxidation process consumes all the dissolved oxygen in the water body causing death of the remaining aquatic life (including fishes) en masse. This phenomenon is scientifically termed **eutrophication** which is a serious matter of environmental concern. But it happens due to other reasons as well. We shall visit them afterwards.

Pesticides and chemical fertilizers are together responsible for another catastrophe in aquatic life. The near-extinction of indigenous, small, fresh-water fish species (mostly cat fishes) is the case. Traditionally, these fishes used to move out of water bodies to roam in paddy fields during monsoon, their breeding season. This would facilitate their breeding and population rise. The latter used to make up population loss due to fishing. But presence of insecticides and chemical fertilizers in the shallow water of paddy fields beyond tolerance limits for those fishes in recent time has caused their mass killing and inhibition of their breeding process. That is why the availability of these fish varieties has become very scarce in markets today.

By far the most grave instance of agrarian water pollution is **arsenic in groundwater**. Although the root-cause is natural–traced back to arsenic-bearing rock strata – it gets accentuated due to excessive extraction of groundwater for irrigation with the help of deep tube-wells. Let me elaborate a little. There are some arsenic minerals in rock strata, basically in the form of arsenides mixed with sulphides. At places they are concentrated enough to form mineral deposits. If the occurrence is located in the catchment area of a major river basin, the minerals are ground to silt and deposited along with alluvium over the floodplains downstream. In fact, this is the case with Ganga River Basin. Arsenide (occurring in suphidic pyrite minerals) are insoluble in water. But, as huge quantity of groundwater is drawn up through deep tubewells for irrigation or water supply (it actually happens commonly) the groundwater table lowers down on account of overdtraft (discharge more than recharge) of groundwater. This results in oxidation of arsenide particles into arsenate in the unsaturated zone above water table – an event that has been prevented by groundwater below water table disallowing air to cause oxidation. Arsenates are, however, sparingly soluble in water. Subsequent rain shower, therefore, dissolve them and take along while percolating down to reach groundwater reservoir and pollute the latter with arsenic, a highly toxic element. This is how large parts of Ganga and Brahmaputra plains in W. Bengal, Bangladesh, Assam are suffering from arsenic pollution.

By the way, overdraft of groundwater in coastal area brings about another problem. I mean to say about the problem of **groundwater salinity.** If there is not much groundwater extraction, the freshwater zone in the soil of coastal area lies above the zone of saline water occurring considerably below the sea level. The divider surface is traced to sea level. As and when fresh groundwater is extracted in excess of the amount recharged by rainfall, the salt-water zone gradually encroaches from sea-side into the zone of fresh water upward and landward. As a result saline water is lifted in the wells instead of fresh water.

Chemical fertilizers are responsible for another matter of concern with regard to farming itself. The principal mineral nutrients for plants are nitrogen, phosphorus and potash (NPK) present in soil. But any or all of them may be deficient in the soil of a farmland. These macro-nutrients are available in chemical fertilizers which are applied to soil in order to make up the deficit and enhance crop yield. In addition to the macro-nutrients, plants need some micro-nutrients like calcium, magnesium, sulphur, zinc, etc. A wise farmer should apply appropriate kinds of fertilizer in correct amount for a given crop according to the soil characteristics determined beforehand. But, in a populous country like India, the uneducated or half-educated farmers make rampant use of chemical fertilizer without testing soil, under pressure for quick increase of food production. As a result, the soil soon loses its natural fertility and productivity, not to speak of the soil pollution due to toxic insecticides.

Animal farming and pisciculture also cause pollution basically of soil and water. It is generally originated from excreta of cattles and fowls which may liberate methane gas into air. Methane is generated from the decay of agricultural farm waste as well. Specially notable is the water pollution caused by prawn-shrimp farms due to various feeds and medicines applied to the ponds from which water needs to be frequently pumped out.

That is all about environment pollution in the farm sector in short. What about other activities of mankind? So let us see how **domestic, institutional** (office, school – college, etc.) and **commercial** (market, shops, etc.) **activities** lead to environmental pollution. The common domestic waste that we see all the time at home is garbage and rubbish. This type of waste collectively generated in a city or town is termed **municipal solid waste (MSW)**. It is generated and collected from homes, offices, educational and other institutions, hotels & restaurants, shops and markets, etc. The wastes from all these sources are more or less of similar nature and comprise wastes of vegetables, fruits, fish-meat-egg, spoilt or waste food; plastic or other carry bags, packagings, bottles, pouches, etc.; waste papers, rags, cartons, boxes, etc.; discarded furniture and utensil; discarded electrical and electronic gadgets like bulb, tube, cassettes, CDs etc; rubbish and rubbles and so on. All of them obviously pollute the environment in this way :-

One : A lion's share of MSW is of putrescent nature made of vegetable and other food waste that putrefy in no time and spreads foul odour in air as well as pathogens through flies. This in turn helps soil and water pollution.

Two : Those which do not putrefy but get decayed to disappear after a considerable time, i.e., biodegradable wastes like rags, waste paper, card board packagings, woodens, metals, straw, rubbish, etc. cause aesthetic pollution on the spot until biodegraded.

Three : Plastics constitute a major portion of garbage on MSW today. This may seem to be a very simple or innocent matter. But actually this is a grave problem especially when plastic carry bags, covers, jars, bottles, cups, pouches and sachets are littered around. It needs no mention that this is quite a commonplace. Because of many practical advantages the use of plastics and polymers in our daily life today is very extensive, growing day by day. The advantages are : absolute impermeability (it is air and water proof), chemical inertness in normal temperature, flexibility and mouldability, transparency, light weight, non-conductance of heat and electricity. On top of that it is very cheap and easily available. But any way, its rampant use has reached such a height that we don't mind to take a bag while going out for shopping. The carry bags from shops make do if it, not to speak of bottles, jars, pouches, sachets and cups so commonly available and used. This scenario is not at all confined to urban life now. It has reached countryside as well.

The irony is that impermeability – the very attribute which has made plastic so popular – has but made it a big enemy of environment. This fact, however, does not usually occur to us. And this ignorance is catalyzed by extreme chemical inertness (at normal temperature) of plastic which has made it suitable for packaging food and medicine. Moreover, we hardly think of environmental pollution other than chemical pollution or microbial infection. The point we can't get at is that the waste plastics, if not collected and disposed off properly, get littered over soil, farmland, forests, ponds, lakes, rivers, sea-shore or along road-side and in drains. This may develop, over a long period, an impervious layer resulting from continued accumulation (For this is not a biodegradable substance). This situation may result in deterioration of quality and productivity of soil and water bodies to such an extent that some problem may arise in regeneration of farm, forest or aquatic resource and, above all, in maintenance of ecological balance. The cause is the impervious plastic barrier in the path of bio-geochemical cycles which account for cyclic regeneration of natural resource. Our awareness and seriousness about this is very essential but lacking actually. Water-logging of roads in cities due to plastic-clogged drains during monsoon – a very usual event – may be an eye-opener in this regard.

Let us now look at discharge of waste water or liquid effluent. **Domestic effluent / waste water** (including that from institutional and commercial places) comprises mainly sewage from toilets / bathrooms and waste water from kitchens, flowing down the sewerage in urban areas. In large cities it is collected at a central

Plate- 2 Plastic litter over a picnic spot on the water-front of a river-dam reservoir at Maithon, Jharkhand

place and treated before being discharged into natural drainage like rivers or streams. In rural or semi-urban areas, domestic effluent makes way to open field, soil or water bodies. On flowing into natural water bodies or other sources of water supply it makes the water polluted. The chief pollutants are suspended solids (TSS), BOD, pH, coliform bacteria (causing enteric diseases) and nutrients (causing entrophication).

The sources of suspended solids are food waste, stool and dirt from bathing, washing, scrubbing, floor cleaning, street sweeping, rain-wash, etc. This pollutant gets intimately mixed with waste water and makes the latter more turbid. There may be other ramifications as well. For example, suspended dirt particles may form shelter of pathogenic microbes and harmful chemicals.

In addition to suspended particles there may be organic matter in colloidal or dissolved state. They are of biodegradable character. That means they are decomposed by being biochemically oxidized with dissolved oxygen and microbes and thus account for the BOD load of the waste water. Occurrence of micro-organisms in natural water bodies is quite obvious. Besides, domestic waste water contains a lot of microbes as well. But the source of ready supply of oxygen for the said oxidation is from dissolved oxygen of the water body, where the waste water flows into. So the DO of the water body gets diminished rapidly unless continuously replenished through continued agitation like river current, sea waves, etc. Since such agitation is not usually available in water bodies, the DO content is depleted causing ill-health and death in the aquatic community. There may be some COD load too and that may also be responsible for depletion of dissolved oxygen.

Domestic effluent is generally acidic, but may be alkaline sometimes. Unless the pH is within permissible range it may cause danger in the receiving water body. That sewage is a principal source of pathogen needs no explanation. Therefore, it essentially needs treatment before being discharged into water bodies. Otherwise there may be proliferation of pathogenic microbes. And TSS is the home for microbes as already mentioned. On the other hand, sewage or domestic effluent contains a lot of nutrients and so, on getting into a water body, is liable to cause entrophication (a phenomenon already explained).

This is, in short, the nature of pollution and pollutants owing to domestic effluent. The tolerance limits of these pollutants have been laid down for the sake of environment protection. The limits are to be strictly maintained before discharge of the effluent into receiving water body. The limits are as such : TSS 100 mg/l (if the effluent is discharged into an inland water body) or 600 mg/l (discharging into urban public sewer) or 200 mg/(if discharged on land for irrigation) pH 5.5-9 (in all cases), BOD maximum 30 mg/l (for discharge into inland water bodies) or 350 mg/l (for urban public sewer) or 100 mg/l (when on land for irrigation). This is the permissible standard for domestic effluent in India.

Well, this sector (domestic-commercial-institutional) is responsible for pollution of air as well. The greatest contribution comes from kitchens – domestic or commercial – and vehicle movement. In the countryside commonly used fuels are coal, kerosene, fire-wood and farm-waste. These generate fugitive emission as huge volumes of fume comprising plenty of SPM (dust), carbon monoxide, oxides of nitrogen (NOx) and hydrocarbon along with the main constituent carbon dioxide. All of these are harmful air pollutants. For example, SPM is responsible for allergy and respiratory diseases. NOx is tearful. Carbon dioxide is behind global warming while carbon monoxide is highly toxic. Fuels – coal in particular – has sulphur content in varying degree and so the smoke may have sulphur dioxide which is acidic and noxious. (The impacts of these pollutants are to be discussed at length afterwards). In this context, cooking gas (LPG), widely used in urban households and restaurants in recent time replacing other traditional fuels, has greatly reduced air pollution.

Now I would tell about a little-known area of institutional environment pollution from the healthcare services. It is of serious concern even as it has not yet attracted our attention enough. Hospitals, nursing homes, polyclinics and diagnostic laboratories generate a lot of waste called **biomedical waste.** The latter consists of many objects responsible for spreading pathogens as well as toxicity. That may, in turn, cause outbreak of disease and even epidemic. Take for example antibiotic vials, injections and syringes, infected dressing, blood-urine-stool samples, etc. Unless properly disposed, but thrown away, the environment including water, air, ground surface or soil may get polluted and unhygienic.

One more thing at last. Being crazy with the fun of crackers and fireworks during Diwali or other festivals, we just don't care about how much air pollution they give rise to. Coupled with too much noise (not to speak of loud-speakers playing music in too high volume), crackers and firewroks emit too much SPM and noxious gases. Playing with chemical colours and "abir-gulals" during Holi is no less pollution. It pollutes water as well as air apart from directly making harm to body. Immersion of idols after Puja also results in water pollution in rivers or other water bodies. For, idols have on them a lot of paints and chemicals of toxic nature and the bamboo-straw structure may rot in water. Tourism may also be responsible for pollution even strongly enough. Well, you can't deny that huge garbage accumulates – or rather is littered – at a tourists' spot on account of excessive crowd. Nowadays plastics, the bio-nondegradable, are found galore in the garbage. On top of that, many spots of tourism happen to be ecologically sensitive such as sea shore, protected forests or national parks.

After so much about pollution in domestic – commercial – institutional sector, we shall now focus on **industrial pollution,** by far the most important of all. The importance of this pollution lies in the volume, extent and diversity of wastes and pollutants. These are much greater here than in other sectors and the impacts of pollution are much more severe. Let me include mining, construction & urbanization and transport (surface, sea and air) in the industrial sector. The activity attracting our attention at the first glance in this sector is combustion of fuels. Fossil fuels are burnt in aircrafts, vehicles, rail engines, ships, machines, furnace, boilers and thermal power stations. Huge quantity of smoke is emitted through exhausts, chimneys or stacks and dispersed in the atmosphere. The smoke or fume consists of SPM, carbon monoxide (CO), sulphur dioxide (SO_2), oxides of nitrogen (NOx) and hydrocarbons along with carbon dioxide (CO_2), the main constituent. If cleaner fuel (say LPG or LNG) or improved technology is used, generation of these pollutants – especially SPM, CO and hydrocarbons — are appreciably reduced. These three pollutants are due to incomplete combustion.

Metal ores (especially base metals), coal and oil generally contains sulphur. Therefore metal smelting, oil refining and combustion of coal and oil generate some SO_2 to get into ambient air. This gas itself is acidic and, being further oxidized in air and coming in contact with water or moisture make sulphurous and sulphuric acid. So where there is considerable emission of SO_2, it makes rain water acidic and brings about **acid rain.** Buildings and other properties made of marbles and metals are corroded and damaged by acid rains or acidic moisture in these places. A notable instance is Agra. As large number of metal foundries are situated in this region and there is a large oil refinery located a Mathura, not far from Agra, the acid rain due to these industries has endangered Tajmahal, the world-famous marble architecture. Transportation industries as well as power plants also emit this gas and causes air pollution similarly.

Burning of fossil fuels give rise to NOx side by side with SO_2. It is produced by oxidation of nitrogen of air through combustion at very high temperature. This is a noxious gas causing eye burn, nausea, headache, etc. Tear gas is known to consist of this. But what is more important is that it is responsible for formation of smog out of fog and smoke. Oxides of nitrogen and unburnt hydrocarbons photochemically react to form thick haze around suspended dust particles. This is termed **photochemical smog,** usually found during dry winter morning.

As I said, combustion in a confined environment with restricted supply of oxygen happens to be incomplete and give birth to lots of carbon monoxide (CO), black (carbon) SPM and residual hydrocarbons all of which are dispersed and pollute air. It occurs mainly in inefficient engines, furnaces or boilers. The modern fuel-efficient ones with improved technology greatly reduces the menace. Carbon monoxide is a highly toxic gas. Well, there are some other sources for emission of CO and hydrocarbons. A notable illustration is underground coal mine where these two gases wreak havoc. Spontaneous but incomplete slow oxidation of coal in the confined environment of underground mines results in generation of CO which accumulates in the same confined environment. On the other hand, hydrocarbons (chiefly methane) is embedded or trapped in coal deposit itself to be released into the mine void and accumulated in the closed void. Methane is an inflammable and potentially explosive gas. Both CO and methane need to be continuously pumped out of the mine void and fresh air is to be ventilated into the same void for safety of the mine workers as well as the mine itself.

The most predominant industrial pollution of air, however, is SPM (suspended particulate matter) in terms of quantity and extent of impact. Apart from stacks and chimneys of factories and vehicle exhausts, SPM also comes from mines, mineral processing or beneficiation and transportation, cement plants, stone quarries and crushers, earth moving work / land development or other construction activities. Out of all these, the case of coal-fired power plants is worth special mention. Fine ash particles blow out of stacks/chimneys of the power plants and is dispersed to a distance of several kilometers flying in air. So this is called **fly ash.** A major portion of fly ash is now-a-days arrested in electrostatic precipitators fitted with the stacks. The coarser fraction of ash remains in the furnace or bottom of stack as residue and so is called **bottom ash.** Coal ash of power plants is a big headache in India. Ash in dry condition, if dumped in open, becomes easily air-borne to from SPM. This is particularly true of fly ash because of its fine grain size. It, therefore, needs to be disposed off in ash ponds as slurry by mixing with water. The problem still persists. Indian coals generally have too much ash content. As a result, huge volume of ash is produced in power sector thereby requiring large ash ponds. In a populous country like India, where land is a scarce commodity, to arrange land in large scale for the ash ponds of power plants is indeed a herculean task. But anyway, SPM in air is a significant air pollutant causing a lot of harm to human health. It is responsible for fog and smog, reduction of visibility, accumulation of dust everywhere and all that. We shall deliberate at length afterwards.

Many more air pollutants (other than SPM) are generated in the industrial sector. Before closing on the topic I shall mention the gaseous pollutants. They may be either **toxic** or otherwise hazardous. Such gases are known to be discharged from industries such as chemicals & petrochemicals, fertilizers, insecticides, paints, drugs & pharmaceuticals, etc. Unless emission of these gaseous pollutants are prevented or arrested carefully to remain within definite permissible limits or, in the event of any accidental / uncontrolled release of these gases, there may be devastating mass casuality in the community living around the factory site. On top of that it may bring about a number of adverse impacts upon the environment at large – air, water flora and fauna. I hope, many of us are aware of the dreadful accident in the pesticide factory of the then multinational company named Union Carbide located at Bhopal that occurred on the 3rd December 1984. Thousands died instantaneously. Many more died afterwards suffering tremendously. Those who survived are languishing even now carrying forward the nightmare with themselves. The accident happened with sudden debouch of the deadly gas called "mic" (methyl isocynate). Thus there is always the risk of several air pollution caused by gas leakage from such industries due to technical fault or mere accident.

Last but not the least to be mentioned in this episode is a gaseous pollutant known as **"Freon".** The chief of this group of halogenated hydrocarbons is chloro-fluorocarbon (CFC). This is used as refrigerant gas in the compressors of fridge and air conditioner. Despite adequate precaution this gas leaks into air, albeit very slowly. The gas ozone is present in the atmosphere at traces. But it is fairly concentrated in the upper layer of atmosphere at a height of about 25 km. This ozone-rich layer engulfing the lower atmosphere is very thin and fragile. But anyway, it carries out a noble cause in nature. Sun rays consist of the killer UV (ultra-violet) rays which is deleterious to the health of mankind along with flora and fauna. But the UV rays are filtered from sun while the latter passes through the ozone layer and thus can't make harm. Freon gases of the atmosphere, on reaching the ozone layer, destroy the latter at places to make holes which enable ultraviolet rays to reach the ground. Such holes have been found above the polar regions of earth.

We now come over to the next episode, i.e., **industrial pollution** – the pollutants and pollution of water environment. The most predominant cause of water pollution is liquid effluent – waste water in most cases – though gaseous or solid wastes may be the cause as well. So, for that matter, we should first characterize the liquid effluent. Production of different industrial products gives rise to various sorts of liquid waste in accordance with specific production process or technology. Let us try to understand this point with the help of a few cases of industry. The group of industry that breaks out water pollution much less than others comprises **primary industries,** i.e., those meant for extraction of natural resources directly. Take for example : mines for extraction of minerals, stone or sand quarries, salt factories, plantations and workshops producing tea-coffee-spices, saw mills, brick kiln, ice factories and so on. To these we may add construction activities such as housing, roads, railways, bridges, townships, ports, airports, industrial estates and plants, so on and so forth. The chance of water pollution is relatively less in these areas. Because the activities here are just extraction and re-placement of natural resources with some handling and marginal make-over. But never-the-less one pollutant is generated profusely. That is **total suspended solid (TSS)** – the dust and dirt particles or sediment suspended in water. This pollutant makes the water turbid and visibly ugly and the suspended particles making sediment are supposed to precipitate at last, thereby silting up rivers, reservoirs or any other water body. The soil and rock fragments dumped after excavation at mining or construction sites are washed with run-off rainwater to be deposited on beds of rivers or other water bodies including dam reservoirs. Well, the menace of TSS in natural water bodies is trifle, for it settle down when the water becomes calm after monsoon. The turbidity goes away. Yet the problem of sediment deposition, as I said just now, still remains. Let me explain. In many cases – mine sites are located in and around river valleys. Interestingly enough, those very regions happen to be the locations of river valley projects with dams and reservoirs. Take for example Damodar Valley, Rehand, Hirakud on Mahanadi, etc. By virtue of easy availability of coal and water, coal-based thermal power plants cluster around the location. Sometimes steel plants, sponge iron and cement plants come up as well. Enormous volume of sediment washed out of quarries and waste dumps of the coal mines as well as from ash ponds/mounds of the power stations and waste dumps of steel, sponge iron or cement plants, is deposited on the bed of dam reservoirs during rainy season. In this way, most of the river-dam reservoirs are getting silted up rapidly and losing their water-storage capacity. The matter, therefore, must be viewed seriously in this light.

Chemical pollution of water caused by mines is also a big menace. The problems are varied for different mines. For instance, in sulphur ore (e.g. pyrite) and base metal mines (having sulphide ore) as well as some coal mines (especially those of northeast India having high-sulphur coal), the water drained out of the mines is appreciably acidic. The deleterious effects of acidic water need no explanation. This event is called **acid mine drainage.** It makes the receiving water bodies acidic as well. Sulphur of coal or ore gets oxidized, on being exposed, and then being dissolved in water forms sulphuric and sulphurous acids to make the water acidic. Allied problems arise out of phosphorite (the mineral raw material for phosphatic fertilizers), fluorite or fluorspar (the mineral used in aluminum metallurgy and responsible for fluoride pollution) and chromite (chromium ore used in stainless steel making as well as tanning industry and causing hexavalent chromium pollution) mines.

Now look at the solid wastes of primary industry sector before concluding on the latter. In case of mines – opencast mines in particular – the principal wastes are mine spoils (often called overburden) comprising soil, rock fragments and waste minerals which are inevitably excavated along with product minerals. Initially they are dumped outside the mine in large heaps and, after sufficient advancement of the mine quarry, dumped in the abandoned quarry to backfill the same. Mine spoils may, sometimes, bear some harmful material such as sulphur / sulphides, fluoride, arsenic, etc. that may cause pollution. Apart from mining, similar pollution through similar wastes may also result from construction activities. In both the cases, huge amount of sediment is likely to be transported into river bed, reservoirs or other water bodies to silt them up.

Now comes the turn of **primary processing** sector which comprises industries involving processing or preparing the natural resources after extraction through simple physical or chemical methods, or both, in order to produce a value-added product. The degree of environmental pollution here is, but, a little more than in primary sector. Take for instance agro-industries such as : rice mill, flower mill, sugar mill, oil mill, jute mill, etc.; simple food processing like dairy, slaughter house, jam-jelly-pickle making, bakery and so on; mineral beneficiation like washeries, crushers or handling plants, coke ovens; cement & ceramics; brick and lime kilns; so on and so forth. An important group of industry i.e., thermal power plants may also be tagged to this sector.

The chief air pollutant in this sector of industry is suspended particulate matter (SPM) which is often accompanied with some gaseous pollutants and odour menace. Barring this there is not much air pollution. But we must not forget that dust and fly ash, debouched from power plants and cement-lime-stone works, is a matter of grave concern in terms of the difficulty faced in their proper disposal and the chances of air pollution otherwise. The extent of water pollution is, however, quite considerable in this sector. The predominant water pollutants are : total suspended solids (TSS), pH, BOD, COD and some others which are industry-specific. There are industry-specific solid wastes as well. As we have already learnt how these common pollutants lead to environmental pollution we shall not go into details on that.

Let us now focus on **secondary – tertiary processing** and **manufacturing** sectors. As a result of immense physical alteration and intense chemical reactions such industries give birth to a great deal of harmful solid wastes, liquid effluents and gaseous emissions. At times these may even be of hazardous nature. This group of industry consists of chemicals & fertilizers, paint & dye, synthetic fibres & textiles, paper, rubber-plastic-polymer, hyde tanning, metallurgical & engineering industries including sponge iron, cosmetics & toileteries, electricals & electronic hardware industries, among others. Almost all kinds of pollutants come out of these industries to a great extent. The chance of pollution in air, water, soil/land, flora & fauna is very high and mitigation is difficult unless a stringent system of monitoring, prevention and control is in place. As the nature of pollution is varied, being industry-specific, let us skip off a detailed deliberation.

Oil well drilling, extraction of oil & natural gas, oil refineries, petrochemicals & lubricant industries, ship-breaking workshops, drugs & pharmaceuticals, pesticides & herbicides and some other manufacturing industries generate large amount of various **hazardous wastes** – in solid, liquid or semi-solid states. In case some gaseous waste of hazardous nature is produced it ought to be arrested within the plant in order to prevent any emission. Before we go ahead with further deliberation it needs to be clarified what exactly is meant by the term "hazardous". Basically, the materials or wastes which are of explosive, highly inflammable, toxic or chemically hyper-reactive nature are identified as hazardous. Because there may be severe damages to lives and properties of mankind or to any ecosystem due to such substance. Bhopal gas tragedy of 1984 needs no explanation to demonstrate this fact. The threat of hazardous waste or substance occurs not only at the sites of industries enumerated above, but also in transportation and handling of the same. Thus oil and gas are vulnerable to leakage not only from oil wells, refineries and petrochem plants but also from transportation pipelines and tankers or storage installation as well. Oil & grease and gas are as highly inflammable – sometime even explosive – as toxic. They contain heavy metals like lead, murcury, chromium, cadmium and non-metals like sulphur, phenol, – all of toxic character. Floating layer of oil & grease over water is a commonplace. This is an indication of water pollution by oil & grease which is liable to cause great harm to the health of man and other creatures including fish. Accidental oil spills over seas and oceans during transportation in ships are frequently heard of in recent years. We can easily guess about the extent of marine water pollution from the hue and cry raised after such events. It threatens the lives of fish and other marine organisms. That petroleum gas is toxic and explosive has been experienced through household accidents which are not so rare. Equally hazardous is any kind of waste moving out of oil refineries, lube or petrochem plants since that too actually has all those hazardous constituents of oil. Many drugs and pharmaceuticals as well as all insecticides and herbicides are toxic. These, along with the intermediate products in their manufacturing process, may accidentally leak out of the plants. This is immaterial whether the accident occurs owing to defective control system for gaseous emission, effluent discharge or solid waste disposal, or just by way of accident. The hazards are all the same.

There was a pesticide factory, owned by an American multi-national company named Union Carbide Corporation, at Bhopal, the capital city of Madhya Pradesh. A number of insecticide brands used to be manufactured there. A highly poisonous gas called "mic" (methyl isocyanate) used to be generated midway in the production process. On the third December in the year 1984, the deadly gas accidentally erupted out through an explosion and got dispersed in the ambient environment. The habitations around the plant, mostly slums, fell prey to it along with the workers within the factory. This killed thousands of people instantaneously. Thousands became seriously sick out of which many succumbed to death afterwards.

Many survivors turned paralyzed. Some of them are still alive carrying forward the nightmare.

By the by, I should tell about the hazards of batteries and e-wastes. The batteries used in various cars, machines, uninterrupted power system (UPS), etc. are made of lead and acid both of which are hazardous. Therefore, if we throw away spent batteries the result may be dangerous. This is the case with wet (lead-acid) batteries. What about dry-cell batteries? Dry cells are used in many appliances such as radio sets, music systems, mobile phones, lap-tops, torches, cameras, etc. These cells are made of toxic heavy metals like nickel, cadmium or chemicals like manganese oxide. Moreover, many other parts of electronic gadgets and computers are composed of toxic heavy metals. Like adding fuel to fire, electronic goods become quickly obsolete, on account of fast scientific and technological advancement, thereby generating heaps of **electronic waste (e-waste).** Disposal of e-waste has thus become a headache in modern society despite all the efforts to reduce the volume of such hazardous wastes by recycling them into new products. But how far the efforts have gone towards success is yet to established. The menace is further worsened because of the plastic bodies encasing these equipments.

Before concluding the brief deliberation on industrial pollution I shall mention a very significant pollutant causing very serious pollution of environment. It is **radioactive pollution.** I hope, many of us are aware of nuclear bombs and nuclear energy. It is based on the premise of an enigmatic property of certain minerals and the very heavy metals extracted from them like uranium, plutonium, radium, thorium, etc. By virtue of this property these materials emit radio-active radiation which inflicts deadly impact upon any living body depending on degree of radio-activity and duration of exposure. The effect may be anything between paralysis or other uncurable disease and death. Besides the plight of people caused by Hirosima-Nagasaki atom-bomb-induced radiation – not to speak of the mass killing – the accidental radiation unleashed from Chernobyl atomic power plant of the then Soviet Union is worth mention. It wreaked great havoc killing thousands of people and injuring or paralyzing many. It was all due to the radioactive emission from the ill-fated plant.

As the last case of anthropogenic pollution of environment, I would like to cite **pollution caused by wars and accidents.** By this I mean the usual events other than the unique tragedies of Hirosime-Nagasaki, Chernobyl or Bhopal. Conventional warfare involves firing, shelling, bombing (with fire broken out therefrom) and excessive movements of tanks, aircrafts, rockets, etc. These activities do cause appreciable air pollution by way of fuel combustion, ammunition firing and explosion-generated gaseous emission adding SPM and noxious gases into the ambient air. Chemical or biological warfare – though banned now through international treaty – needs no explanation. The impacts are obviously understood from the very names themselves. Common accidental cases like air crash, fire break-out, etc. are also responsible for air pollution.

And that is all about how the environment gets polluted by various natural causes and much more by the developmental activities of mankind. Now we should move on to see the adverse impacts inflicted upon the health and properties of man along with the entire biosphere by environmental pollutions and how to prevent, control and mitigate them. So we come to that point in the next chapter.

5
Pollution of Environment–Impacts & Mitigation

Let us spell out at the outset – pollution of environment means presence of pollutant (one or more) in a given environment. Assessment of pollution – whether in air, water, soil or land – should, therefore, be in accordance with the significant pollutants of each environment one by one. Dealing with air first, let us consider pollution by **suspended particulate matter (SPM),** the simplest and most common air pollutant. It is by far the greatest in terms of quantity as well. Suspended particles finer than 10 microns, called **respirable particulate matter (RPM)** enter through nostrils into the respiratory system including lungs. This gives rise to various respiratory diseases such as cough, allergy, asthma, bronchitis, tuberculosis and the likes. Stone particles flying out of stone crushers or opencast mines may cause silicosis. Asbestos particles from mines or processing units are liable for cancer. Smokes – especially tobacco smoke inhaled through direct or passive smoking–is also a probable cause of lung cancer. Moreover, a major constituent of smoke is carbon-rich SPM. So it is essential to keep a tab on SPM and smoke in order to allow within tolerance limit as per **National Ambient Air Quality Standard (NAAQS).** For that we should control emission and prevent generation of dust and smoke at the source itself as far as practicable. But how?

Smoke and dust are erupted from factories or furnaces through chimney or stack and then dispersed in air. A device known as **electrostatic precipitator (ESP)** is, therefore, fitted in the stack or chimney in order to arrest SPM before emission into atmosphere. This is very effectively and widely used in thermal power stations to control the menace of fly ash. There are some other types of devices, e.g., bag filter, baffle, cyclone separator, scrubber and all that in order to collect dust and reduce smoke. This is how smoke and dust generated at a fixed location is controlled or prevented. But the task becomes all the more difficult when the emission is of fugitive character (occurring here and there over an area in a scattered manner). **Fugitive emission** is characteristic of mines (opencast mines in particular), stone crushers, coal and mineral handling plants / beneficiation plants, transportation of coal or other minerals and overburden / mine spoils, etc. The most simple and successful mechanism for prevention and control of fugitive dust is water sprinkling over the sources of dust in a fine shower as far as possible. Movement of vehicle traffic also generates dust and smoke from road as well as from exhausts. The best practice for mitigation is to install fuel – efficient engines and improved exhaust systems along with their regular up-keep and maintenance.

Oxides of nitrogen (NOx), occurring in smoke, exceeding the tolerance limit locally may cause tears, headache, etc. It is known as a basic component of tear gas. That apart, NOx chemically reacts with other pollutant like hydrocarbons in sun and on mixing with SPM of the smoke form what is known as **photochemical smog.** This brings about haze in the atmosphere drastically diminishing visibility. That in turn leads to many hazards like disruption of rail and air traffic and human sickness as well. Emission of NOx is prevented easier than controlled. Since its origin is combustion at very high temperature it can be considerably reduced by technology upgradation of engines or furnaces. Otherwise, some absorption mechanisms may be applied.

We have already learnt that burning fossil fuels obviously generates **sulphur dioxide (SO_2).** The gas is acidic, toxic and pungent. The origin is sulphur content in the fuels. SO_2 is sparingly soluble in water to produce unstable sulphurous acid. But it may be oxidized in air and turned into sulphur trioxide that, being dissolved in water, produces strong and stable sulphuric acid. Supply of water in atmosphere readily comes as rain shower, dew or fog. SO_2 thus turns rain shower, dew of fog acidic. This event is known as **acid rain.** The adverse effects of acid rain is quite obvious. Buildings, furniture and equipments made of marble, metal or even common cement may be corroded if exposed to acid rain or SO_2 emission. Many heritage monuments such as Taj Mahal and Victoria Memorial are thus being damaged day by day. Besides, forests, agriculture, soil, water bodies, flora and fauna are also likely to be damaged owing to destabilization of biochemical pH balance in the ecosystem. NOx may also be responsible for acid rain to some extent. Prevention of SO_2 emission is possible with use of low-sulphur fuels. For example, LPG and LNG (petroleum gas products) are now being used in place of coal and diesel oil. The stacks or chimneys of power plants or other industries may be made higher enabling the emission to disperse more and get diluted. In iron & steel or other metallurgical plants, oil refineries and coke ovens, SO_2 is converted to sulphuric acid to be recovered as a byproduct.

Carbon monoxide (CO) is a highly toxic gas. Very slight rise of its presence in air make it cross the tolerance limit leading to fatal result. As soon as this gas reaches blood of man or any other creature through respiration, serious illness occurs. Further increase of CO content leads to death. Since incomplete combustion in a confined environment is the cause, it may be avoided by making the fuel combustion more efficient with oxygen rich environment in furnaces or engines and by applying improved technology. Good arrangement for fire prevention and fire extinguishing as well as improved ventilation system (for example, in underground mine or basement of buildings) also help in this direction. For, adequate supply of oxygen and well ventilated environment help generation of carbon dioxide instead of carbon monoxide.

Carbon dioxide (CO_2) is produced through complete combustion of organic substance – be it fossil fuel or biomass – through innumerable events and activities. It is added to atmosphere through many other means as well. Chiefly for example, respiration of organisms, decay of dead bodies, volcanic eruption from the earth's crust, forest fire, coal mine fire or any other fire occurrence. Recalling the carbon cycle, CO_2 is captured back from atmosphere into biosphere, hydrosphere and lithosphere through various means. These are basically photosynthesis by plant and green algae as well as absorption by oceanic water to create limestone strata on sea bed. There exists a harmonized system of balance between emission of CO_2 and its sequestration in nature. This accounts for the stable proportion of CO_2 in the gas mixture of air maintained for ecological balance until the same got disturbed chiefly by human activities. The latter came in through rapid industrial development on earth ushered in by Industrial Revolution. This resulted in extensive and voluminous production of CO_2 and its emission into atmosphere through industrial and allied activities and consequent increase of its proportion in air. The trend didn't stop there. It is continuing perhaps at a faster rate now. CO_2, being present in air in a very small amount (0.03%), is mildly acidic, though seemingly inert and innocent in that sense. It has no bad impact upon health instantly. But, on account of development activities of mankind, its ever-increasing proportion in atmosphere has a very detrimental impact upon environment globally. Let me explain, in short, how.

The environment in most parts of Europe and North America is very cold. Horticulture there is, therefore, next to impossible during winter. In order to protect the tender saplings and herbs from severe cold, a kind of "green house" is built in nurseries there. This is nothing but a closed shed made of transparent glass cover. Sunlight, available only for a short while, warm up the ambience within the green house. The warmth is retained in the shed till late night while the environment in the surrounding is quickly chilled early in the evening. This is made possible by carbon dioxide and water vapour released by plants (through respiration) and soil moisture (through evaporation) getting concentrated in the confined environment of green house. The heat obtained from sunlight during day time is trapped by these two gases (CO_2 and water vapour) to be retained till late night in green houses while the heat outside is lost by quick radiation. This phenomenon is comparable with a cloudy night which is found warmer than usual owing to higher humidity – the water vapour trapping heat. Carbon dioxide performs the same job in our atmosphere. The earth's surface get heated by sun during day-time and radiates back the heat to atmosphere from afternoon to midnight. Gases like CO_2, NOx, water vapour, methane and other hydrocarbons occurring in atmosphere trap a portion of the radiated heat for sometime. As a result, ambient air is not so quickly cooled as is the earth's surface. It remains warmer for sometime, similarly as the air inside a green house, compared to the ambient air. This phenomenon is known as **greenhouse effect** in allusion to the green house (earlier cited) and the gases responsible for this phenomenon (just named) are termed **greenhouse gases.** Interestingly enough although the phenomenon has been responsible for proliferation of life on earth by warming up the atmosphere and the hydrosphere according to the majority of scientific community, it is threatening to cause disaster in future. Let me come to that point now.

Concentration of water vapour in air varies with the humour of nature (weather) but is beyond control of man. But the other greenhouse gases are being added to atmosphere every day by human activities ranging from farming through domestic fuel burning, deforestation, transportation, power generation to commerce and industry. Well, any gas in free state cannot stay stationed at a place but gets diffused until it is homogeneously distributed in the entire atmosphere. Therefore, given a case of excessive emission of a greenhouse gas at a certain location, it is dispersed to raise the concentration of the gas all over the globe however small may be the rise. As a result, the entire atmosphere is warmed up gradually over a long period. The event is called **global warming** which in turn leads to **climate change.** It means, under the impact of ever-growing greenhouse gases the atmosphere over the globe is very slowly but gradually warming up resulting in the possibility of erratic behaviour of climate. For example, arid or semi-arid regions may sometimes receive high rainfall whereas high-rainfall zones may face drought. Otherwise, colder regions may become warmer and hot belts may turn cold. There comes an erratic behaviour of climate on the whole.

Some scientists have recently gathered evidence, through the results of different surveys and research projects, that the atmosphere and the hydrosphere are getting warmer very slowly but steadily. At the same time, the warming process is being accelerated as well. In the last two decades or three, the temperature has risen up by about half a degree Celsius and the rise may be upto two degrees or three after a century. This may look negligible. But no, just think of the far-reaching consequences. It is already found that the famous twin glaciers of the Himalayas – Gangotree and Jamunotree, the origins of the two premier rivers of India –

have conspicuously shrunk over decades. The great polar ice caps of our planet have had the same fate. Have you ever thought of the dreadful ramifications likely to occur at different parts of the globe (which may even be accelerated by ever-growing greenhouse gas generation) ? Melting of ice caps will add huge volume of water to the oceans. That, in turn, will lead to rising sea level and consequent flood along coastline submerging extensive low-lying coastal area. There is no dearth of such vulnerable places. Good examples are – Sundarbans, Lakshadweep Islands, the Netharlands and many more. Moreover, the perturbed water cycle due to climate change may bring in unusual drought and flood causing great damage to farm and forest resources. It may even lead to scarcity of water flow in the Himalayan rivers like Ganga, Jamuna, Indus, Brahmaputra – the lifelines of Indian people – the impacts being obvious. What is the solution then ? That is very difficult as the cause is almost inevitable and the impact is global. So the mitigation needs international effort about which I shall state at some length later on. Now I would just mention that in order to control carbon emission (responsible for greenhouse effect) we should practise minimum energy and fuel consumption, fuel-efficient and energy-efficient technology and renewable or non-conventional energy application.

As I said, the ozone layer enveloping our lower atmosphere is getting damaged by another group of gases called **freon** or halogenated hydrocarbons emanating from industry and household. These gases make **hole across the shield of ozone layer** while the shield protects the biosphere from the deleterious effects of ultraviolet rays of sun. On being exposed to ultra-vioilet rays easily passing through the said holes there may be skin burn, even skin cancer. Delicate lives may even be killed. Mind it that the evil impact of freon does not confine itself locally but goes global as in case of greenhouse gas. Otherwise why do we find a big hole in the ozone layer over a polar country Antarctica ? Fortunately enough, there is hardly any population there. Yet this is a matter of grave concern. For who knows if such a hole is going to appear over a populous country, say India for example, in future ?

Many more kinds of gaseous pollutants are discharged from different industries. Or, they may be generated as intermediate products that may accidentally leak out owing to mechanical failure. Many of them may be toxic. Various types of pollution take place in different industries depending upon the product and manufacturing process. Thus, industrial sectors like oil refinery & petrochemicals, fertilizers & chemicals, cosmetics and pharmaceuticals, insecticides, plastics & polymers, synthetic fibres, metallurgy, food processing, etc. liberate different kinds of gaseous pollutants – toxic or otherwise harmful. Unless adequate precaution, monitoring and firm preventive measures are in place, sudden emission of such gases is likely to play Bhopal-like havoc for the workers as well as the people around.

An almost unavoidable property of modern society is **noise.** This is particularly characteristic of urban and industrial area where unpleasant sound (called noise) spreads from vehicle traffic, factories, workshops, machines, crackers, loud speakers and crowd. Sound is a form of energy that, being generated from any kind of vibration or movement of objects, propagates in the form of wave through a medium. In case of noise in our environment the medium is air. So noise is a kind of air pollution being the unpleasant or irritating form of sound in contrast with music (instrumental or vocal) or normal conversation which are pleasant or comfortable. If a man is placed in the ambience of loud noise, for some time, he or she falls sick depending upon the degree of noise and exposure duration. The same applies to other creatures as well. The consequence may be deafness, headache, mental depression, unrest, hypertension and all that. Apart from these, it is well known that noise disturbs sleep. I have already said about noise level parameters and tolerance limits. Stress of noise beyond tolerance limits make body and mind sick. Remember, animals are much more sensitive to noise than is man. We should, therefore, strictly avoid creation of noise and, at least, keep it within tolerance limit.

After air pollution we come over to see water pollution – water body and flow on the surface and groundwater. What calls our attention at the very outset is **turbidity** which in turn owes to **total suspended solids (TSS)** present in water. Suspended dirt particles cause many water-borne disease like cholera (now known to be eradicated in India), diarrhoea, gastro-enteritis, dysentery, typhoid, jaundice, etc. Dirty stagnant water, being the breeding house of mosquito, is responsible for outbreak of vector borne disease like dengue, malaria, kala-ajar, etc. Colour and odour, if anyhow come in, make the water absolutely unsuitable for domestic use. Turbidity may be removed from water by allowing to settle in a sedimentation tank with, if required, applying alum and then filtering depending upon use. Waste water discharged from many industries remains hot and so should be cooled down before letting into receiving water body.

As I said, natural water is supposed to be neither acidic nor alkaline. But owing to many reasons–application of chemical fertilizers in the fields or waste water from various industries including mines – **pH** of surface water or groundwater may be raised or lowered. This may give rise to many problems in water environment with regard to aquatic life and domestic water in particular. Another menace arises from **water hardness**, found much more in groundwater than in surface water. **Hard water** takes too much soap or detergent to wash clothes and on boiling or storage forms crust on the boiler or the pot. **Iron** dissolved in water in a very small dose is good for health at least for anaemic persons and expecting mothers. But if the concentration exceeds a given limit of standard many ill impacts come in, such as constipation problem and rust-staining on containers and utensils. Iron too is more common in groundwater than in surface water just like hardness. I think, this has been experienced by big housing complexes in the outskirts of metro cities where supply is sourced from deep tube-wells for want of municipal supply. Hardness may be removed simply by boiling and then filtering out the carbonate precipitate. There are special techniques for removal of other specific salts including those responsible for iron menace. As iron is present in water as dissolved ferrous salts (which later gets oxidized to be deposited as brown precipitate or stains of ferric oxides and hydroxides) it is required to be driven out by reverse osmosis (RO) or some chemical methods. A number of **RO filters** have come up in markets today even for domestic use. This is capable of removing both iron and other dissolved salts causing hardness. Acidity in water can be neutralized by applying lime. Similarly, alkalinity may be driven out with limited application of weak and harmless acid. Fly ash is being introduced now to control pH.

Fluoride pollution, as I said, occurs in surface or groundwater owing either geochemically to strata or to mining and some other specific industries like aluminum and iron smelting. If present in water exceeding tolerance limit it may cause health problems chiefly in bones or teeth. A disease known as fluorosis is caused by fluoride pollution even as fluoride is applied in small dose for dental care.

Taking cue from strata-generated pollution, I come to a very serious matter of concern. I mean arsenic pollution, a topic of recent hue and cry. How it happens is already stated. Let us now hear about its impact and mitigation. Consumption of arsenic-contaminated water brings about a lot of health problems. Arsenic is highly toxic and carcinogenic as well. That is why its tolerance limit is very low in all the water quality standards. Prolonged ingestion of arsenic beyond tolerance limit may cause skin cancer. Sometimes it may even lead to death. Outbreak of arsenic pollution is spectacular over lower Indo-gangetic plains and Brahmaputra plain. Almost the entire plains of Bangladesh, Gangetic West Bengal, Assam and Sindh of Pakistan are reeling under its impact. Currently some sort of **arsenic filters** have come up. They have been put to use to some extent with government and public efforts. But I can't say how far they have been effective. If we revisit the discussion regarding origin of arsenic pollution, we can realize that preventive measures taken effectively would be more fruitful. For, despite the root cause being natural (geological) arsenic pollution is enhanced by excessive extraction of groundwater through deep tubewells. We should therefore minimize the practice – be it for irrigation or for water supply, if not to stop altogether. Instead, use if surface water should be stressed upon. To comply with that, groundwater recharge is to be facilitated more and more with extensive practice of rainwater harvesting.

We have already learnt about water pollution due to fertilizers and pesticides in the farm sector. Water source meant for drinking or other domestic use, if intoxicated with insecticide beyond tolerance limit, is obviously harmful to health. But the point of more concern is that continued ingestion of this poison, even in a very small dose, for a prolonged period may lead to incremental accumulation within the body. This phenomenon, termed **bio-**

accumulation may turn the situation from simple illness to a fatal one. Today we are no more ignorant of carcinogenic impact of toxic bio-accumulation. On the other hand, the ramifications of toxic water pollution may enter into the food chains of an ecosystem. Let me illustrate in short. Suppose, pesticide enters into the body of fishes in a pond. The same poison later goes into a man's body through eating the said fish. Through another food chain, some cattle feed on pesticide-polluted grass and straw of a farm. The said man feeds on milk and meat of affected cattle. Thus pesticide finds way into the man's body through two food chains thereby ultimately magnifying the concentration of the poison many times. This event is called **bio-magnification.** To deal with the dual menace of bio-accumulation and bio-magnification of toxic pollution we should avoid or minimize pesticide application. Only pesticides of natural organic formulation, like Neem, or short-lived chemical ones may be prescribed. Those of persistent character is not at all advisable. Farming with natural pesticides and organic manure instead of chemical pesticides and fertilizers – called **organic farming**–is an emerging idea and should be welcome. Toxicity of water environment may come from industry as well and the impacts are likewise. Prevention and control measures vary in accordance with production process and technology. Deliberation at length on that is not relevant here.

The chemical character (e.g. pH) of water around farmland may change owing to fertilizers. There is no doubt about it. But, in addition, another important hazard may come in. that is **eutrophication** in water bodies due to nutrients of fertilizers. Eutrophication may be caused by other sources of nutrients like food processing industries, distilleries, dairies and slaughter houses as well as sewage. How it happens has been explained earlier. The impact is sudden famine of dissolved oxygen in the affected water body and consequent massacre of aquatic lives like fishes. Mitigation is possible only through prevention.

BOD and **COD** are also responsible for depletion of dissolved oxygen (DO) in water, albeit not as severaly as eutrophication. So the adverse impacts are similar but and not so severe. But never the less it cannot be ignored as it 'for sure' affects the health of aquatic organisms. So removal of these two pollutants from effluents before flowing into water body is very important. Here again the best solution is prevention of generation of the pollutants. However, BOD and COD of waste water are required to be brought down, by treatment, below tolerance limits as per prescribed standard. Then the treated water should be recycled into the industrial use instead of discharging out. We have seen that DO in water is consumed for biochemical oxidation of organic matter (which accounts for BOD) eaten up by microbes present in the sludge at bottom. In doing so, DO content of the water body is gradually diminished. This can be countered by steadily supplying DO into the water by artificial means like whirling or agitating the water through some device, say, fan. Flowing water naturally does the job. Such things are commonly found in prawn farms where water is agitated and periodically refreshed as shrimps and prawns are very DO-sensitive. But anyway, industrial effluent is generally required to be treated in treatment plants for removal of BOD and COD alongwith other pollutants.

A simple effluent treatment plant has a primary setting tank where sludge and suspended dirt particles (TSS) is mostly settled and removed. The effluent then flows into a large shallow tank called aeration tank where a powerful fan continuously rotates in the upper layer of water to agitate and inject air into the water. This keeps on supplying DO for biochemical oxidation of organic impurities by microbes rising up from the sludge at bottom. The microbes – otherwise dormant in the sludge–become very active. They then quickly devour organics (BOD) in the water and oxidize the same to make neutral sludge that settles at the bottom of the tank. The sludge at once move up and become activated again by the agitation. That is why the sludge is called activated sludge and the treatment process termed **activated sludge process**. BOD-free water is now shifted to another tank where a slow-moving fan gathers the floating sludge into flocks to settle sown at the bottom of the tank. The water above is then transferred into a secondary settling tank where turbidity is removed from the water with alum dosing. After that, the clear water is stored in a clear water reservoir or recycled into use. If required the water is disinfected with a chlorine dose especially for domestic use.

Let us now look at the scenario of land or soil pollution. Just keep in mind that all soluble pollutants pollute soil through groundwater and so the resultant pollution is of similar nature with similar impacts. But there may be many insoluble pollutants in the form of solid wastes as well. In addition to different types of polluting wastes dumped from various industries, the following are worth special mention : ash from coal-fired power stations, slag from iron & steel plants or other metallurgical industries, soil & rock waste (called over burden / gangue / rejects) out of mines, rejects from mineral beneficiation plants, paper pulp waste and bulk wastes from other sundry industries. There is, on the other hand, municipal solid waste comprising garbage and rubbish from various source. These undoubtedly gives rise to **aesthetic pollution of land** and, on top of that, propagate pathogenic and chemical contamination. The wastes, therefore must be disposed off in a proper manner. But then, the problem remains with the fact that disposal of these wastes needs vast areas of land, a very scarce commodity in a populous country like India. Adding fuel to fire the problem is further aggravated by high ash-content of Indian coals. The result is voluminous generation of **ash** in coal-based power plants. About 80% of the ash fly up through stack and most

Plate 3 : Utilization of power-plant ash in a coal mine of Talcher Coalfield, Orrisa, for back-filling of a decoaled quarry (ash slurry in the foreground and the quarry in the background).

of the fly ash is collected in ESP in order to prevent dispersal in air to increase SPM level in the ambient air. The rest 20% bottom ash, being coarser, remains at bottom and removed from the plant along with fly ash collected in ESP, for disposal. Coal ash (fly ash in particular), being readily responsible for dust pollution, is not fit for open dumping on ground. So it is commonly disposed off as water-mixed slurry in ash ponds. But that requires a lot of land which, as I said, is scarcely available in India. Ash disposal has thus become a big headache for power sector. So we should think of means to reuse the ash as a resource instead of disposing off as waste. And that is possible in various ways. For instance, given fly ash of suitable character (physically and chemically), it may be utilized as a raw material in manufacture of pozolona cement as well as bricks and blocks for building construction. We can also make use of fly, bottom or pond ash in land-filling for road, rail-road, bridge, dam or any other construction. We can even back-fill a mine voide (especially open-pit mines) with ash from pit-head power station. This is an emerging concept of land reclamation in open-cast mines called "ash haul-back." Besides, ash may be utilized in many sundry jobs like soil treatment or treatment of waste water as well.

Slag, a bulk waste emerging out of iron & steel and other metallurgical plants, is responsible for soil or land pollution apart from requiring much land for disposal. As it is usually sulphur - bearing it is likely to make soil acidic. Moreover, prolonged accumulation in bulk makes heaps of slag looking like hills around steel plants or sponge iron plants. But fortunately the chemical composition of the slag of steel plants makes it suitable as a raw material for cement, more so than fly ash. Slag is, therefore now being extensively utilized in cement manufacture. That is why we don't find as much big heap of slag today as would find during yester-years around a steel plant. There is, however, a large scope of using this waste as road metal.

Municipal Solid Waste (MSW), also generated in bulk, is another menace of serious concern. Unless disposed properly it may cause land pollution through spreading disease apart from causing ugly aesthetics and putrid odour. Since this waste is of heterogenous character, it should be segregated into different fractions that need different disposal practices. For instance, recycleable objects like paper, metals, glassware, plastics, etc. are to be sorted out first and collectively sent for **recycling** to respective destination. Similarly, putrid and easily biodegradable materials such as waste food, vegetable waste, etc. may be separated and sent for **composting or biogas generation**. An emerging idea is to make use of such biomass in generation of electricity. The remaining third portion – chiefly rubbish short of things – may be disposed in land filling hygienically, the method being termed **sanitary land-fill**. Hazardous material, if any, should be sorted and put to incineration. It is very urgent to prohibit plastic littering. We must not forget it.

We have known about pollution due to electronic, hazardous, biomedical and radioactive wastes. E-wastes are best taken care of by recycling. Some electronic hardware companies are now taking back obsolete and discarded material for recycling. Hazardous wastes need a special disposal method called **secured land-fill**, i.e., a land-fill system preventing leaking or leaching of hazardous waste through surface runoff or groundwater percolation. Any small leachate coming out is moved back to the disposal pit. For biomedical wastes there are several sophisticated techniques and equipments like incinerators and autoclaves. We need not go into details. We should only keep in mind that they should be kept separately for proper disposal rather than mixed up with common garbage. Radioactive wastes needs a very sophisticated and fool-proof mechanism of disposal.

Before I conclude the topic of land or soil pollution and impacts, I should make out a very relevant point–**change of land use pattern**. Land use evolution due to urbanization and industrialization with evolution of human society has already been talked about. This involves farmland turning into residential plots or industrial compound; establishment of factory or residential complex after filling up water body or wetland; township replacing villages or farmland; clearing forest for constructing industry, river dam & reservoir, mining or resettlement of ousted villagers by developing fresh farmland and homestream land. In addition to such change of use pattern there may be transformation of the basic characterisitics of land such as topography, drainage pattern, soil, greenery and so on. The most spectacular illustration can be found in case of mining. In an open-pit mine, mineral is raised after extensive excavation of land to remove rock and soil cover over the deposit. In case there is a forest, that is to be cleared first. If there be a village instead, that needs to be shifted to a resettlement site. After mining, an artificial large pit or quarry is formed at the site along with large hill-like dumps of discarded mixture of soil and fragmented rock outside. Top soil – very essential for growing trees, shrubs and herbs – is lost in this operation along with forest, village or farmland. Natural drainage system lost and replaced by man-made alternative may give rise to many ecological problems. Thus the entire environmental scenario undergoes metamorphosis. The scenario of an underground mine on the other hand is somewhat different. Extracting mineral by making long and wide tunnels through a mineral deposit results in voids underground and consequent subsidence of land as is seen in many parts of Raniganj and Jharia coalfields. Extensive occurrence of fire in Jharia coalfield is a national disaster. Immediate re-location of villages and most of Jharia town itself is essential but is not forthcoming. This fire is also responsible for collapse of houses. Besides, the

Plate 4 : Biological reclamation of mined-out land with plantation over spoil dumps of an open-cast coal mine.

groundwater table goes down quickly because of continous dewatering of mine by pumping out groundwater. The following steps are required in solution to mine-related land problem. In case of open-cast mines the land should be restored to premining condition as far as practicable. At least the degraded land should be reclaimed by backfilling the quarries with the excavated rock dobris (instead of dumping outside) immediately after mining and then covered with a layer of top soil initially excavated from the place. Then it should be further reclaimed biologically with planting diverse plant species, preferably indegenous. This activity is called **land reclamation**, widely practised in Indian mines. As for underground mines, the solution to land subsidence is to fill in the abandoned voids with sand pushed into as slurry with water. This prevents subsidence in a big way.

So we have made a brief idea about the impact of air, water and land/soil. By the way, what about the impact upon the biosphere ? As I said, biosphere extends over atmosphere, hydrosphere and lithosphere particularly across their interfaces. Life processes run being supported by free and uninterrupted flow of matter and energy among the four arms of environment. So the biosphere is obviously affected due to pollution along with air, water and land thereby disturbing the delicate ecological balance. As a result, populations of different organic species is endangered. Some species may even face exinction. This is coupled with direct mass-killing of many species by mankind. Massive reduction of forest area, dwindling wildlife with endangered species of rehino, lion, peacock, tortoise, olive ridley sea turtles, etc. and famine of hilsa (the farvourite fish of Bengalee people) due to overfishing may be cited as glaring examples.

Through the deliberation this far we saw how environment gets polluted, what are the impacts and how to mitigate in different areas of human activity. But it is all the more important to tackle the problem holistically in an integrated manner. Then how to do that ? I suggest a three-pronged solution. In the first place, the problem should be arrested before coming in. That means prevention of pollution. We should adopt best practices in various developmental operations (industrial production, agriculture, mining, construction services and so on) in such a way that wastes or pollutants would be generated at the bare minimum. Always keep in mind the basic principle : "prevention is better than cure" which is very much appropriate in case of environmental protection. The other two solutions, i.e. control of pollution and mitigation of impact are secondary to that. They may only be alternative or additional measures. Therefore, we ought to make a proper, well-thought out and detailed plan for this objective before we go in for any developmental activity. An essential feature of this plan should be application of modern, environment-friendly technology. Another essential component of the plan is prediction about the nature and degree of pollution likely to occur owing to the proposed developmental activity. This activity is termed **environmental impact assessment (EIA)**.

So how do we do that ? The first step required is to study at length the existing environmental scenario of air, water, soil & land use, flora & fauna before initiation of the project implementation. This is known as **basclinc survey/monitoring.** We have already learnt about monitoring of environmental quality parameters. This survey is done basically through such monitoring at baseline phase (before execution) with respect to each significant parameters. In the second step, a detailed assessment is to be done about probable impact of each developmental activity upon each arm of environment. Normally this is done with production of an **EIA document**. The next action is to prepare a judicious integrated plan for prevention, control and mitigation of probable pollution or any other adverse impact upon environment. This plan is termed **environmental management plan (EMP)**. Finally this plan must be implemented hand-in-hand with execution of the project. Once again I would remind that the first priority should be given to prevention, secondly to control and lastly to remedial mitigation measures. In case all the three options appear to fail, the proposed project should be outrightly rejected. Any industry or other developmental project is bound to be associated with pollution of environment. But it is possible to keep tab on pollution or damage to environment by reasonably acting in accordance with a correct EIA/EMP. Therefore it is very important that an EIA-EMP document should prescribe best production process and technology so as to minimize generation or liberation of wastes or pollutants.

Over and above, there should be provision for measures to restrict pollution within tolerance limit. If required further, there may also be provision of remedial measures.

6

Our Morrows – Environment Protection & Sustainable Development

Let me recapitulate a little. Man—the species Homo sapiens evolved from wild animals in a primitive natural environment. In course of hunting in forest for animal flesh and fruit-root to subsist in the said environment mankind learnt agriculture, animal breeding and poultry farming, wearing and dwelling in huts. Gradually developed a social system, evolution of society and growing civilization with time. Then came in organized population, habitat in the form of villages, improved agro-animal farming, improved huts or houses, roads and carriage, education, health-awareness and medicare system, crude or cottage industry, commerce & trade and all that. In this process mankind has achieved uplift of the quality of life.

Thus in modern time we have got all the amenities like modern cities and towns, building & furniture-fittings, improved roads and modern vehicles, railways, shipping and air transport, tele-communication and computers with internet, processed food, fashionable dress, jewellery, cosmetics, advanced education and healthcare, modern commerce and above all, industrial enterprises. All these have been possible through modernization of traditional agriculture and animal-poultry farming coupled with extensive industrialization and urbanisation, ushered in by the Industrial Revolution in Europe. Our society has been developing in this way. But then, all of these activities leading to elevation of quality of our life need natural resources in adequate quantity. These resources are basically sourced only from environment which, in plain terms, means air, water, land/soil and the living world along with the energy inherent in them.

Natural resources required for development are, however, not uniformly available everywhere. They are best available over plains, especially along river banks and coastline. Let me give a few reasons. The best land suitable for ago-animal farming is available on river flood-plains or coastal plains where the alluvial soil is fertile and water-bearing. Construction of houses and roads is more convenient over here. Realization of this fact has made much more rapid growth of human settlements over here, relatively. Out there in hilly, plateau or arid regions we find scanty habitation. Because of port facility along river banks or sea shore commercial opportunity is much more available here. For the same reason and for easy availability of water, skilled man-power (because of established habitations like town and cities), better communication and infrastructure facilities these locations are at much more advantage than others for establishment of industry as well. But there are a few riders. For instance, mines may come up in a barren desert land or a plateau or a hilly region if enough mineral deposit is found to occur. The mines are followed by mineral-based downstream industry including power plant. Or, say, hill stations like Jammu & Kashmir, Himachal, Uttarakhand, Darjeeling and the North-East where tourism industry has developed. Anyway, centered around such prospective locations, called **growth centres**, socio-economic development, industrialization, urbanization and land-use evolution take place in course of time.

By the way, how is land utilised, by man ? The most primitive and natural land use is forest and water bodies on which early man used to subsist. These two are basically integral parts of most ecosystems as well. Through social evolution of mankind, the basic land uses such as residential, agriculture, animal & poultry farming, pasture, ponds & pisci-culture, crematory & cemetery, market, etc. came in. Gradually afterwards, in course of extensive urbanization, industrialization and socio-economic development, land has been extensively used for towns & cities, roads & railways, ports & airports, irrigation & drainage, river valley projects (with large dams, reservoirs and hydro-power), industrial establishment, commercial farms, plantation, and other developmental activities. On the other hand, population is growing exponentially. What is the consequence then ? The pressure of ever-growing demand for land is coming from either sides–traditional use as well as modern developmental use. The pressure is all the more intense in a developing populous nation like ours. The scenario is very conspicuous in the thickly populated states like West Bengal and Kerala.

So how do we meet the fast-growing land demand? Land is not an infinite resource. It is true at least for usable land even as it does not seemingly occur to us. So the hunger for land is met by invading into forest land, farm land or water bodies. Even homestead land is not spared and villagers are ousted for developmental project. And that becomes unavoidable in case of site-specific projects like mining, port, river dam, etc. For it is not feasible to shift them to alternative sites for technical reason. For example, a mine can only be opened at the site where exactly the mineral deposit occurs. On the other hand, people displaced by a project would certainly look for resettlement site, leave aside the question of rehabilitation of their livelihood. Similarly farmers would for sure need alternate farmland on losing the same in order to subsist. That is why we find land losers sometimes encroaching into forest land and carve out alternate farmland. Sometimes they fill up public wetland to develop homestead. Moreover, under pressure for development, industrialization and urbanization take place by directly erasing forest and filling up wetland. Then where does it lead to? Shrinkage of farm land and water body/wetland for now and loss of forest land for ever. But we know for sure that forest and water bodies play essential roles in environment or biosphere. And dwindling farmland is not at all desirable for the sake of food security in a populous developing country like India. So, in my opinion, this fact should be kept in mind during clearance of a project and acquisition of land for the same.

Back to the point, I would cite a few more instances. Take the case of Sundarban. Historically, this has been a classic example of vast ecosystem spread over the world's largest delta with dense mangrove forest and estuarine backwater to support a rich biodiversity. Please take note of the "World Heritage Site" status recently conferred on it. Look at its present fate on the ground. Apart from being politically partitioned between India and Bangladesh, large chunks of mangrove forest and back-water wetland in the upstream region have been eaten up by expanding Greater Kolkata metropolis and associated industrial belt. The fate of the fabulous Amazon Rain Forests having the richest terrestrial biodivesrity on earth is just comparable. Singrauli region of Madhya Pradesh (Sidhi District bordering Uttar Pradesh)–where earlier there was dense forest over rocky plateau–is now replete with large open-cast coal mines, several mega power plants, Rihand Dam river valley project together with associated townships and ancillary industries. There are so many comparable instances like Durgapur-Asansol, Dhanbad-Bokaro, Talcher-Angul (Orissa), Chandrapur (Maharastra) industrial belts, to name a few. In all the cases large areas of forest or farmland have been ravaged to turn into mined out derelict land, industrial township and plants with heaps of slag, ash or other wastes. At the same time many villages have been vacated for mining or industry, or submerged for river valley project ousting millions of villagers and making them lose their livelihood. The river valley projects submerge vast areas including ecologically valuable forests along with farmland. On the other hand, port-based growth of industrialization and urbanization along river bank or sea-shore is very common all over the world. Indian metros like Mumbai, Chennai, Kolkata, Kochi, etc. are glaring examples. There have been development at the cost of ecologically valuable mangrove forests and wetland or farmland. Anyway, the thing is that there has been continued change or evolution of land use or land envionment because of industrialization and urbanization all over the globe.

Thus it is established that we primarily need plenty of land of various nature along with water or water bodies for various activities of socio-economic development with a view to elevate the quality of our life. At the same time we require many other natural resources such as forest resources like wood, bamboo, seeds, honey, etc.; various mineral resources like coal, oil & gas, metal ores, stones, sand & gravel, etc.; and various sources of energy such as thermal, wind, solar, atomic, hydro-power, etc, Besides, there are so many man-made resources useful for our development. For example farm products and processed food, agroindustrial products, other manufactured products, buildings & other construction, infrastrcture and various services. Yet the root source of all the manmade resources are nothing but natural resources available only in environment or nature. Furthermore, all the flora and fauna from mankind to micro-organisms get all the vital ingredients–water, air with oxygen, food & nutrition, habitat and everything–from none other than environment. There is no alternative. This unique characteristic of environment, i.e., the capacity to support life process is called **supportive capacity**.

While elevating our quality of life more and more through the developmental activities stated above by utilizing various natural resources, a lot of wastes along with pollutants are generated inevitably. This we have already learnt. We have also known that, not only from developmental activities, but also from our day-to-day domestic activities, many wastes and pollutants do come in. Even other creatures do the same by way of excretion. These wastes and pullutants–anthopogenic and natural–would have progressively degraded the enviroument or nature and made the latter unfit for supporting life unless, through the very own processes of nature, the wastes and pollutants would disappear gradually. It happens through (bio-) degradation, physical disintegration, chemical decomposition and transformation, dilution, etc. This unique capacity of environment (nature) to absorb into itself and make disappear the wastes and pollutants is termed **assimilative capacity**. Isn't it something like Lord Shiva's gulping voluminous poison erupted from the "Samudra Manthan" and turning "Nil-Kantha" (blue-neck) in order to save the earth?

The above two terms (supportive capacity and assimilative capacity) being clubbed together make the term **carrying capacity**. The carrying process of the earth goes on relentlessly for ever. And this is possible through a large number of bio-geochemical cycles, earlier told about.

It is, therefore, very urgent to ensure the stability of these cycles in order to secure the carrying capacity of every ecosystem for protection of the entire biosphere required to sustain continued quality addition to our life. Otherwise, the socio-economic development, running fast in our society, may come to a halt some day rendering our future generations to face a dark unbearable life. Because, there would be very little resources available for development, and not at all a good environment for healthy life either. All that would be available are polluted air, inadequate and polluted water, limited derelict land, scarce fuel, scanty energy and (due to extinction of many species) ecosystems starved of biodiversity.

What is the solution then? We should lead life and carry out development in such a careful manner now, that there is no dearth of resources in our environment–in quality or quantity–to be left for development sustained through generations to come. Such a system of development may be called **sustainable development**– a term much talked about but, perhaps, not-seriously felt for. The essential requirement for this are protection of environment and conservation of natural resources in terms of quality as well as quantity. But how to do so? I should prescribe the follwing:

1. In the run-up to any developmental project or programme there should be a well-thought out plan in place to make sure that the natural resources–land, water, air, energy, forest or any other ecosystem including flora and fauna therein, human habitat, etc.– are judiciously utilized in optimum. Special care is to be taken to ensure minimum change in traditional pattern of land use. This should also be guaranteed about what is said in the plan is implemented properly during execution of the project. Rampant use of natural resources–be it land (forest, farmland or wetland in particular) or water (of surface or groundwater regime)–with an eye on instant profit should be restrained. This point is very important while selecting the site of an industry.

2. We should allow application of only the right kinds of fertilizers and pesticides as much as is justified, in the agro sector. At the same time, we should arrest rampant groundwater overdraught through deep tube-wells in order to maintain groundwater balance. Shallow wells and ponds are always a better option for irrigation.

3. In the industrial sector, improved process and technology should be adopted so as to achieve maximum possible output of product with minimum input of raw material, water, fuel and energy. Simultaneously we should see to it that minimum waste is generated and that too biodegradable rather than persistent. In short, only eco- friendly technology is advisable.

4. Disposal of wastes in proper manner so that it does not have much adverse impact upon environment. If warranted, the wastes are to be treated enough before disposal. Hazardous wastes must be disposed in a specially designed secured manner, more so in case of radioactive wastes. Use of plastic should be restricted

at the bare minimum and stringent prohibition should be enforced upon littering in biodivesity-rich or ecologically fragile environment as well as at countryside.

5. Production of byproduct should be aimed at in place of waste. Otherwise effort should be in place to reuse or recycle wastes as much as practicable, after treatment if required.

6. Extensive utilization of renewable energy and resources (such as hydro-power, solar energy, wind power, ethanol, bio-diesel, biomass, etc.) to the maximum. Concurrently, use of non-renewable and finite resources like fossil fuels and minerals to be kept at the bare minimum with enough conservation for future.

7. It is very essential to maintain biodiversity and ecological balance in every ecosystem. Marginal damage may be accepted for unavoidable reason. But there must be mitigation of the damage and rehabilitation. As experts say, at least one third of the total land area ought to be covered with forest. But in most countries including India it is far below. We should, therefore, try to make up by planting more and more trees and reclaiming degraded forests.

8. In arid, semi-arid or water-shed regions, we should preserve existing water bodies and add new ones to harvest rain water more and more. Similarly, in urban areas, especially in housing complexes, rain-water harvesting is to be done by having tanks on roof-tops and pushing the harvested water into recharge wells. This may help groundwater recharge and maintenance of water table.

9. Effluent discharge and emission into air from all sources must be abated. All the pollution parameters must be kept within tolerance limits. Continuous monitoring of air emission, ambient air, effluent discharge and receiving water bodies is required for the purpose, along with treatment warranted by the monitoring.

10. Environmental quality monitoring at every sources of pollution and the affected areas around must be done all the time. As and when any failure is detected, effective prevention and control measures are to be taken accordingly.

11. Prior to execution of any development project (industrial or otherwise) an appropriate environmental impact assessment and an environmental management plan (EIA - EMP) must be rolled out. Implementation of the EIA - EMP during execution of the project is to be ensured as well.
12. Without going into further details (not possible here, either) I would suggest that a national environmental policy and regulatory system should consider the above recommendations. The system, however is existing in many countries including ours. But the status of implementation is usually found poor. In fact, the task is herculean because of social, economic, political and administrative reasons.
13. Strong campaign for environmental awareness and extensive propagation among public at large are urgently needed along with strengthening of sincerity and responsibility among people as well as the government.
14. It is essential, above all, to keep tab on population in order for environment protection. This is my view and, I am afraid, this view may not be shared by many. Every man (as well as any other creature) is a consumer of environmental resources and at the same time producer of pollutant–both for life process (applicable for all) and development (applicable for man only). In other words, he or she is responsible for depletion of natural resources (at least the non-renewable ones) and enhancement of pollution. So our environment is liable to be deteriorated in proportion with rise of population density as overall availability of environmental resources may be considered to be proportional with land area in a simplified model. The carrying capacity of a geographical region may, thus, roughly be correlated with population density.

It is a bare fact that actually at the root of global problems like food crisis, water scarcity and global warming lie the causes like explosive population growth, unjust land use, excessive or inappropriate application of chemical fertilizer and insecticides, too much extraction of groundwater, unplanned urbanization and uncontrolled industrial pollution. Population has nearly reached explosive stage in countries like India and Bangladesh. I believe that sustainable development is next to impossible side by side with uncontrolled population. I remember, in the first world summit on environment held at stockholm in 1972, Indira Gandhi, the then premier of India, was compelled to say that India's first priority was poverty alleviation rather than environment protection.

To demonstrate the pressure of population upon environment let me consider the case of food security in the first place. In India and Bangladesh, farmland is split into tiny plots on account of population density. Owing to food crisis it is imperative to quickly augment food grain production in limited small plots. In this backdrop and on the event of famine-like situation in the nineteen sixties, India underwent "Green Revolution" through extensive application of chemical fertilizers, pesticides, irrigation and artificially evolved high-yielding seed varieties. This, albeit solving food crises, brought in many problems as well. What are those? An irony occurred due to unfit and unbalanced application of fertilizers and pesticides, overdraft of groundwater and cultivation of new high-yielding varieties of crop species instead of indigenous ones conducive to the environment. Soils have turned infertile and dependent upon chemical fertilizer. Pollination of crops is failing as a result of mass killing of insects. Water bodies are polluted with pesticides and fertilizers and water table lowered down causing thereby arsenic pollution in alluvial plains and groundwater salination in coastal plains. Introduced alien crops being more prone to disease and insect infestation have raised cost on insecticide and fertilizer. Inspite of that our country cannot resort to eco-friendly, traditional or organic farming. On the contrary, India is slowly inclining towards adoption of GM (genetically modified) crops (say, of brinjal and cotton) in disregard of hue and cry raised against these by environmental scientist. Neither has been possible eviction of illegal encroachers in forestland (since they happen to be poor tribals), nor to resist overfishing in seas. Because the nation has the liability to provide food for the huge population as well as habitat and means of living for them. That is why, instead of going in for the difficult task of rehabilitation and resettlement for so many people after eviction from forestland, the nation has legislated the Forest Rights Act, 2006, in order to give the tribals legal right to stay on there and earn livelihood from the minor forest produce.

Exactly for the same reason (population and poverty), this country has not been able to do away with Special Economic Zones (SEZ) –anti-environment and hence banned in developed countries–on the plea of poverty alleviation through fast development.

So we have had massive industrialization and urbanization keeping pace with the fast-growing population to provide the latter with a decent living. As a result, in a populous developing country like ours, change in land use pattern, pollution of environment and resource crunch–all averse to sustainable development–are bound to occur. For, development in agriculture, commerce, industry and services can not be maintained in a judicious and controlled manner under tremendous pressure of meeting the basic needs of the overwhelming population. Besides, there is the urge for further industrialization and urbanization on the same locations of plains or sea coasts (where the environment has already been degraded) because of the practical advantages already cited. Thus it has been historically found that the trend of environmental quality with time in a given geographical entity (say, a country or a state) leads downward (if depicted in a graph) while the trends of living standard of people (quality of life) and (economic) development both run upward, almost parallely. I have tried to illustrate this with a simple diagram given below:

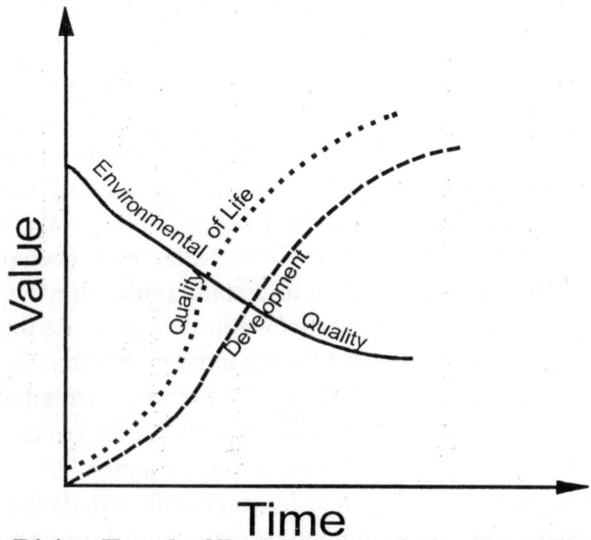

Fig. 8 : Rising Trend of Development & Quality of Life vis-a-vis Falling Trend of Environment Quality with Time

The irony of our fate is that there is always clash between environment on one hand and development (or uplift of quality of life) on the other. This is more so in a developing country where the national priority is inclined towards development rendering the task of environment protection and sustainable development tougher than in a developed country. Another road block is the lack of international understanding for solution of global problems. The causes of certain environmental problems, as I said, are local but the effects are global. Take for instance, global warming, hole in the atmospheric ozone envelope, oceanic pollution and the threatened existence of some organic species like whales, hilsa (sea fish), Olive Ridley turtles, etc. Greenhouse gases are generated locally from power, transport, industry and agriculture sectors. But they get dispersed over the globe in no time and cause warming of atmosphere globally. Similarly ozone-depleting gases are emitted from foam factories and cooling appliances in populated industrial countries. But a big hole has bee found over the virgin continent of Antarctica. Whales are hunted by people of polar region and Hilsa is relished upon only by Indian (Bengalee) and Bangladeshi people. But both the animal groups are endangered over the globe owing to overfishing. Olive Ridley turtles come all the way from the Pacific Ocean to Orrisa coast every year in the breeding season to lay eggs. Because of mass killing of this rare reptile along with eggs over there, existence of the species is now endangered. Similarly a large number of wandering birds fly over to Indian water bodies like Chilka, Sunderbans, etc. during winter every year from far flung territories like Siberia. But, on account of pollution and growing human activities over there, arrival of the birds are diminishing year after year. Summing up, then, we need an appropriate international understanding to sort out these global problems. Because environmental policies and statutes are not uniform for all the nations whereas for the fault of one country some other nation or the whole world may be affected. For tunately, albeit not enough, several understandings have been executed under the aegis of the United Nations. Notable among them are (cited chronologically) :-

1. **Ramsar Convention, 1971** : For conservation of large wetlands and inland water bodies of significant bidiversity or heritage importance. India is a stake-holder by signing and ratifying the document. Several lakes and Wetlands of India such as Chilka, Wooler, East Calcutta (Group of) Wellands (locally called Bheris) are enlisted in it.
2. **UN Conference on Man & Environment, 1972** : The first world summit on environment held at Stockholm. Along with many other nations, India was a party to it. This marked the beginning of environmental awareness and observance of the World Environment Day on the 5th June every year across the world.
3. **Basel Convention, 1989** : Venued at Basel, Europe, this convention formulated the regulations for trans-country transportation of hazardous chemicals and wastes. India is a signatory to it.
4. **Montreal Protocol, 1989** : This is an international understanding reached by many nations (including India) at Montreal, Canada, to evolve the programme and procedure for doing away with ozone-depleting gases.
5. **Rio Earth Summit, 1992** : The second environmental world summit held at Rio-de-Janeiro city of Brazil, under the aegis of the United Nations, on the theme of "**Environment & Development**". Most nations of the world assembled in this summit wherefrom evolved the **UN Framework Convention on Climate Change (UNFCC)**. India is one among the signatory and ratifying countries. The modalities, action plans and responsibilities on the part of developed, developing and under-developed nations towards the goal of controlling greenhouse gas generation were formulated in this summit.
6. **Convention of Biological Diversity, 1992** : In the same year (1992), perhaps as follow-up, this convention was organised aiming at protection of biodiversity over the globe. Our country has put signature in this treaty of pledge and programme to protcct biodiversity and has ratified it in 1994.

7. **Kyoto Protocol, 1997** : This protocol was executed at Kyoto city of Japan in 1997 for regulation of greenhouse gas emission with prescribed timeframes and targets (upto 2012 AD) on the basis of UNFCC, 1992. India, even being a developing country, has signed and ratified to shoulder its international responsibility. But surprisingly enough, the USA being a rich developed country has failed to ratify it after signing. This has put a big question mark on the success for implementation of the protocol.
8. **World Summit on Sustainable Development, 2002** : This has been held in recent past at Johannesburg City of South Africa in order to review the action plans and their implementation status with regard to the protocols of Kyoto, Montreal, etc. and to fix up further programme. But it has not been so fruitful owing to disagreement amongst the nations.
9. **Rio + 20 Summit on Green Economics, 2012** : This world summit under the aegis of UN, venued once again at Rio-de-Janeiro after 20 years, with the theme of **Environmental Economics**. It is an emerging subject to quantify the economic implications of environmental degradation or pollution vis-a-vis economic benefits of environment protection or pollution control.

Even after all these efforts, the status of international understanding towards environment protection is not so encouraging. Thus, global warming, hole in the ozone envelope, dwindling biodiversity, wide-spread oceanic pollution, international trade in hazardous wastes and all those global problems remain far from solution.

Let us tell about another irony–the dilemma between safety of health and protection of environment. By way of health awareness in modern life we need hygienic container, cover, packaging, storage and transportation of food commodities and medicine. Traditionally paper, cloth, hesian, wood, earth, glass or metal used to be used for the purpose. But since the advent of plastics it has replaced them for this job because of a number of practical advantages of this new-found naterial. The advantages are chemical inertness (in normal temperature), absolute air-water-proof property, transparency, flexibility, light weight, insulation to heat and electricity and cheap price. The trend is rising day by day. Hence we commonly find plastic or pet bottles, jar, cup, pouch or sachet in place of traditional glassware, earthware or metalware. Polythene instead of paper, cloth or metal foil for covering or packaging. Jute or cloth bags and paper sacks giving way to nylon/plastic/ polyester sacks, packets or carry bags. Use of disposable syringe to prevent HIV infection. But being charmed with so many practical advantages we often fail to care that plastics are not biodegradable. They are persistent enough to remain in place for thousands of years and so get littered around. Today plastic litter can be found anywhere–road side, drains, open ground, river banks, sea shore, ponds, so on and so forth. But, because of its impervious property and persistence, plastic litter is likely to cause serious environmental problems. Water logging of streets in cities during heavy rains due to drains choked with plastic litter is, I think, known to everybody. I would call attention to a more serious matter, hitherto little known. There may form a pretty layer of plastic litter at a given place through accumulation over years. As we know, the bio-geochemical cycles operate through free and continuous movement of matter and energy among atmosphere, hydroshere, lithosphere and biosphere through their interface. And this movement is obstructed by the impervious layer of plastic litter formed along the interface, just told about. On the other hand, we know that ecosystems earn carrying capacity only through these cycles. So the environment of the place in question is likely to lose carrying capacity. Let me demonstrate this fact with a very simple case. If you look with curious eyes at a spot where the soil is covered with plastic garbage, you find that not even a grass grows over there. Taking cue from injection syringe, not only the disposable syringes, but also waste drugs or empty discarded vials and all the biomedical wastes are liable for environmental pollution unless they are disposed off properly.

The dilemma lies not only with healthcare but also in the area of socio-economics as well. Let me cite a few instances in the back-drop of our own country. As I said, because of easy availability of natural resources, growth centres of socio-economic development are located on plains–preferably along river banks or coastline. In contrast plateau, hills, arid zones, forest or other remote areas remain under-developed. This disparity has historically occurred for obvious reason. But human aspiration doesn't understand this point. So there is growing urge among the people of backward areas to get industries established there. The motive is scope of economic development and employment opportunity. For this reason sometimes industrialists are lured to such backward places with extra benefits and relaxation even under government initiative. Under such relaxation, environmental regulations may also come in. With an eye on commercial profit, entrepreneurs promptly come forward to establish the project there. But what is the outcome ? Rare forest resource, wetland or fertile farmland is lost. Is it desireable in the context of environment protection or food security for a populous country like India ? A judicious decision is required to resolve this dilemma. Recent fiascos of Singur, Nandigram, Posco, Noida are the glaring instances of local people's fierce resistance against acquisition of farmland.

Most of our country's mineral deposits are located in forest or inhabited locations. A very good recent example is the vast bauxite deposit occurring in Gandhamadan Forest of Kalahandi, Orrisa. But the forest is very dense with a rich biodiversity of flora and fauna. Graced by the Government of Orrisa with an intent to exploit the rich resource of aluminium ore for the economic development of the backward region, Vedanta, a multinational owned by a non-resident Indian offered to mine the deposit. As soon as the event came to the notice of environmentalists they, along with the local tribals, put up stiff resistance. The case moved upto the Supreme Court and mining in the forest zone was debarred. As for mineral deposits of inhabited locations, the best examples are Indian coalfields mostly lying over villages and even small towns, e.g., Raniganj, Jharia, Giridih, Talcher, Wardha Velley, etc.

Before I move over to the next point I would like to mention a recent ironical incident. In the year 2003 AD, an NGO named Goa Foundation moved the supreme court of India with a PIL (Public Interest Litigation) case that a large number of industries (including mines) had been operating without having environmental clearance mandated under the EIA (Environmental Impact Assessment) Notification then in vogue. This was viewed as a serious legal offence and the NGO prayed for immediate intervention of the court. The apex court immediately issued closure notices to all the defaulting industrial units. Incidentally, many coal mines (almost all the old ones) found place in the list. The Central Government was shocked like bolt from the blue. The government saw that this would lead to absolute collapse of power generation and supply system across the country for want of coal (the principal source of power) in the event of sudden closure of so many mines at a time. Arguing with this plea, the government prayed to the court for grant of an exceptional exemption with regard to the coal mines allowing for some reasonable time to obtain statutory clearance. The court pragmatically ordered for separate listing of the coal mines and allowed for a few years' time to obtain environmental clearance. But, interestingly enough, some mines are yet to complete the process of clearance after so many years.

And that is all about the irony of the conflict between environment protection and food security, health safety or enhancement of the quality of life. We shall now focus on another very significant topic–the dilemma inherent within the very activities of environmental protection and prevention & control of pollution. It is not possible here to describe all cases. I shall just mention a few sample cases of special significance.

For the sake of prevention and control of pollution we go in for proper disposal or reuse of solid wastes, treatment or purification of effluent discharge or gaseous emission. But sometimes it is paradoxically seen that these very activities may cause pollution. Take for example coal ash of thermal power plants. Since Indian coal is generally of very high - ash grade, coal - fired power stations in India produces huge volumes of ash. The ash is liable to cause considerable pollution (especially fly ash responsible for air pollution). So it is to be disposed off in a suitable manner. Traditionally this task requires ash ponds occupying large area. For a country so populous like India, wasting so much land in such a non-productive use is not at all desirable. In this back-drop, an emerging practice is to utilize the ash in back-filling the de-coaled quarries of the nearby coal mines. But this leads to a probable bad impact upon environment. How is that? Fly ash being composed of very fine particles–and for that matter being very impervious–

may block the path of groundwater recharge. Rainwater cannot percolate down through the ash plug formed in the back-filled quarry. It is creation of another problem to solve one.

Plastic litter is another menace already told about. I would highlight the dilemma in its disposal. The option is either recycling or disposal in a safe manner. Recycling plastic waste more than once is undesirable as that leads to health hazards and environmental pollution as well. On top of that, plastic is a bio-nondegradable or persistent substance. So the only option left is burning. But open burning of plastics gives birth to dioxin, a carcinogenic toxic gas. So burning in the open is not desirable. On the other hand, it is not a simple task to safely burn huge volumes of plastic waste in autoclave or incinerator. Solution of this dilemma is perhaps far away.

Similar dilemma is faced in treatment of waste water. It is often found that disposal of residual wastes out of treatment plants is all the more difficult. Because its toxicity or pollution capability is much more than the original waste water. So the waste residue must be disposed in a very safe and secure manner which may not be achieved easily. Thus it is sometimes found that the environment of the site of disposal of ultimate waste turns highly polluted. Such sites, therefore, must be selected far away from habilitation.

Let me now turn our focus at another dilemma. A present paradigm of our environmental regulatory system envisages divisions of the entire country or each state into "red" (no go), "orange" (buffer) and "green" (go) zones to indicate existing environmental scenario of pollution. Red means highly polluted, leaving little scope for allowing further industrialization. Orange means relatively less polluted having scope of some more industries, but, with stringent regulatory conditions while green is meant for almost unpolluted environment. A very popular word called "**greenfield industry**" is in vogue nowadays. It is based on the conception that establishment of an industry in such an almost unpolluted "green" zone is likely to bring about very little pollution. And, being impressed with this notion, large - scale industrialization is encouraged in such regions. I am afraid to share this view. For, even as the likelihood of pollution exceeding tolerance limits is very less here in the present scenario on account of low base-line values of quality parameters, the latter is bound to jump up after commissioning of the industry. But this thought is overlooked. As a matter of fact, the urge for rigorous environmental management is not felt by the project proponent in view of a better baseline scenario. So the entrepreneur does not take adequate environmental measures. As a result, the environment gets more and more polluted even though the parameters remain within tolerance limits. And in effect, the carrying capacity of the outer environment at large gets diminished. No wonder, it may so happen that a green zone of today may turn orange someday in future. Moreover, you can't rule out the global impacts like greenhouse effect, for example, even if the local impact is not so felt.

Another commonly - heared term and widely encouraged practice is "**eco-tourism**". The objective is to spread environmental knowledge among people. The public at large are being encouraged to visit wildlife sanctuaries, national parks, islands, large water bodies and all those biodiversity-rich fragile ecosystems in the name of eco-tourism. This has, no doubt, brought about enlargement of the crowd of tourism in these eco-sensitive zones. But I have enough reservation about how many of them are really lovers of nature and how much progress has been achieved in enhancing environmental awareness among people in general. On the contrary, the delicate ecological balance is getting disturbed in absence of disciplined environmental habits among the crowd of tourism. Notable examples are Sundarbans, Lakshadweep, Chilka, Dal Lake, Corbett, Kajiranga, Simplipal, Periyar, other national parks and tiger/biosphere reserves. Dal Lake of Kashmir has been found polluted simply by the discharge of Shrinagar City sewerage, thereby largely damaging its original bio-diversity. On top of that, there is the common event of careless throw of garbages (including plastics cup-bottles and carry bags) by the tourists from Sikara. The problem is further aggravated by the stay of tourists in houseboats with direct discharge of sewage. I can't support this to stand by environment. Similarly, there is rampant movement of crude motor-boats in Chilka and Sundarbans spilling oil and lubricants and throwing out plastic carry-bags and cup-bottles. The Sundarbans ecosystem is already ravaged with the fast-growing population of the suburbs of Greater Kolkata. It is needless to say about the damage to wildlife by the menace of picnic and car movement in the wildlife sanctuaries.

Let me wind up the topic of dilemma after citing just one more case. All the fossil fuels including coal, petroleum and natural gas are non-renewable finite resources. Their reserves are very limited in relation to our demand. We, therefore, need to exercise restraint in their consumption and, concurrently, switch over more and more to renewable energy from wind, running water, biofuel and other sources. Thus there have been a great many river-valley projects with large dam-reservoirs for hydel power. But the environment has paid heavily for that. Vast land has been submerged. That includes forestland, farmland and habitations. Lots of people have lost their homestead and livelihoods. Hydropower is, of course, not the only utility of river valley projects. There are multipurpose projects serving other purpose like irrigation, water supply and industry. For instance, Damodar Valley, Bhakra-Nangal, Mahanadi, Cauvery, etc. Having said that, we may stress upon the use of groundwater instead of surface water in order to avoid river valley projects and save homes, land (including forests) and earnings of rural people. But then, this action may invite some other environmental problems such as lowering of ground-water table, salination of groundwater in coastal zones and the fateful arsenic pollution. There are many such instances, but we may not go into further details.

Summing up, sustainable development alone is the solution for elevation of our quality of life continued through generations. And the key to that lies in protection of environment and conservation of natural resources. But there are two tough road-blocks in this direction. One is the inherent clash of development versus environment (more so in developing countries like ours) and the other is the dilemma in the process of environment protection and pollution abatement. We, therefore, ought to know what is to be done keeping in mind the importance of awareness about environment protection cum sustainable development and sincerity in performing the actions prescribed for the purpose. I would suggest the following principal lines of action that may require private and public initiatives :-

1. Environmental studies from the primary level of education and that should be imparted, albeit, in an interesting, simple and lucid manner. It is heartening that there has been some progress in this direction with the intervention of the honourable Supreme Court of India.
2. Wide publicity through mass-communication media–newspapers & magazines, TV & radio, functions and festivals–about the basic environmental concepts and necessity of pollution abatement and conservation of natural resources.
3. Induction and expansion of sincerity in environmental practice and discipline among people are to be made sure. This needs a little explanation. Barring the uneducated ignorant people, I have seen many educated people (who are supposed to be aware enough) throwing away plastic or toxic garbage into drains or water bodies. Such people are, sometimes, even found to be engaged in filling up water bodies with commercial motive. I have further seen such people indulging in felling trees, clearing forests or hunting wild animals or birds. Despite knowing well that plastics are anti-environment we can't give up the bad habit of throwing away waste plastics here and there, not to speak of reducing their use. I have seen beach sands of Marina in Chennai or Swargadwar at Puri–both very famous and always croweded impregnated with plastic cups, bottles, sachets and carry-bags. Similar is the view of waters of Dal Lake and the sea-route from the Gateway of India to Elephanta Island, Mumbai.
4. Although it is an established fact that the best way to spread awareness and sincerity in the society is through education and mass-communication media, the latter have not been found effective enough. That is why we need enactment of laws and formulation of rules along with enforcement of the same through administration and judicial system. Thus there are a number of acts and rules in nations of the world for the sake of environment protection and abatement of pollution. Our country too has passed several acts in the Parliament and thereafter framed and notified a good many rules and notifications. Notable among them are (to mention a few, given in chronological order) :-

i) **Wildlife Protection Act, 1972, and Amendment, 2002 and the Rules** framed thereunder.

ii) **The Water (Prevention & Control of Pollution) Act, 1974 and the Rules** thereunder such as **the Water (PCP) Rules, 1974 and the Water Cess Rules, 1977.**

iii) **Forest Conservation Act, 1980 and Rules, 1981, 2002.**

iv) **The Air (Prevention & Control of Pollution) Act, 1981 and the Air (PCP) Rules, 1981.**

v) **The Environment Protection (EP) Act, 1986** and a number of Rules and Notification framed under that, such as, **Environmental Audit / Environmental Statement Notification, 1992, 2002; Environmental Impact Assessment (EIA) Notification, 1994, 2006, 2009** (the process of environmental clearance, and monitoring of statutory compliance after clearance, by the Central or State Ministry of Environment and Forest, is governed under this notification); **Hazardous Wastes/Substance (Management & Handling) Rules, 1989, 2006; Coastal Regulatory Zones (CRZ) Notification, 1991**; etc. Thus a large number of regulation and notification have been made under this legislation (EP Act, 1986) and so it is popularly called the **Umbrella Act**.

vi) **The Indian - Biodiversity Act, 2002 and Regulations** thereunder.

5. But anyway, legislation of statutes is not the last word. It becomes useless unless enforced appropriately. For that matter we need competent and transparent administration as well as effective judicial system. At present, India is having s Green Bench in the Supreme Court as well as in each High Court to hear cases relating to environmental matter. This is a welcome development. I think, further details on legal provisions may not be desirable here. As for administration, it should be responsible for planning, research & development (R&D), approval/clearance and constant monitoring with right application of the statutes, punitive measures as per law in every case of violation or contravention. The administration should be proactive for propagation of awareness and sincerity among

people as well. Various environmental institutions have been put in place at national and state levels. Notable among them are : The Central Pollution Control Board (CPCB) and the State Pollution Control Board (SPCB) functioning respectively under the central and state Ministry of Environment & Forest (MOEF), National River Commission, National Environmental Engineering Research Institute (NEERI), Wildlife Institute of India, Forest Research Institute (FRI), Renewable Energy Development Authority, Bureau of Energy Efficiency (BEE), Central Groundwater Board and Authority (CGWB and CGWA), etc. Besides, there are universities and research institutes to promote environmental research. All of these organizations contribute efforts to promote protection of environment, pollution control and resource conservation. The efforts are reflected in the establishment of large number of reserve / protected forests, wildlife sanctuaries, national parks, several biosphere reserves and tiger reserves, the Ganga - Jamuna Action Plan, Fly Ash Mission, so on and so forth. The paramount role is, however, played by MOEF at the Centre and SPCBs in the States.

But by far the most important task is dissemination of environmental sense and discipline among the public at large. For this purpose we need, in essence, sincere effort at all levels– personal governmental, non-governmental or social service organization. It is heartening that the United Nations has played the pioneering role in this direction. It organized the first global conference on environment under the banner "Man & Environment" at Stockholm on June 5 of the year 1972 AD. **The World Environment Day** is observed across the world on the 5th June every year since then. It is celebrated in India through various functions and awareness campaigns at government, non-government and community levels. Through the initiative of United Nations' Environment Programme (UNEP) every year a particular theme is selected along with a slogan from amongst different aspects of environment protection in order to be publicized widely all over the globe. As a matter of fact, the whole world had almost been indifferent about environment before this. But there has been found, however, indication about environmental awareness in ancient India through the term "**Pancha Bhuta**". This collective term comprising "Kshiti, Apa, Teja, Marut and Byom" is mentioned in Indian mythology. The five words respectively mean soil, water, energy, air and sky. Barring the word "Byom" (the sky), all the remaining four indicate the three physical elements of environment and the energy inherent in them. The biosphere alone has been left out and the sky invoked instead. Let us leave it anyway. As I said, the interest and sincerity for environment was evolved only after 1972 AD even as the desired level is yet to be reached. Twenty years later, in 1992, the UN once again assembled world nations for another global conference at Rio-de-Janeiro city of South America. It is known as Rio Earth Summit. The theme of this summit was Development & Environment. It is from here that the concept of **Sustainable Development** has taken shape. Again after 20 years (2012 AD) another world summit has taken place in the same city (Rio-de-Janeiro) with the theme of **Green Economics**. It has been named as + 20 Summit on Green Economics, 2012.

Despite all these efforts the success is not that encouraging. We need, therefore, much more effort, intention, sincerity and discipline at every level–personal, community, governmental or institutional. At the same time we need far more strong environmental administration and fool-proof environmental regulation. There are a lot of loopholes in our environmental statutes. The language is ambiguous at many places and that is why it is not possible to take adequate legal action in many cases of contravention. On top of that there are social and political compulsions.

So, for the sake of environment and sustainable development, we need, in essence, sincere concerted effort. Come on, let us save enough natural resources and pollution - free environment for our successor generations. Let us pledge, earnestly and sincerely, to protect our environment in a disciplined manner so as to sustain the flow of development through generations to come, with healthy and prosperous life. Let not the environmental exercise be confined within meetings, seminars and conferences (as is usually seen) but be practically reflected in our regular activities. Come on, let us raise the slogan :

"SAVE TODAY'S ENVIRONMENT

FOR MANY MORROWS' DEVELOPMENT."